The Great Transition

A tale of how it turned out right

Anyone who believes exponential growth can go on forever in a finite world is either a madman or an economist.

Kenneth Boulding, economist

Contents

Foreword

Humanity appears caught in a trap with no way out. 'Business as usual' is no longer an option. However, halting and reversing our consumption of more and more 'stuff' appears likely to trigger a massive depression with serious unemployment and poverty. This is certainly true if all we do is 'apply the brakes' without fundamentally redesigning the whole economic system. We are facing a series of interlinked systemic problems – consuming beyond our planetary limits; untenable inequality; growing economic instability and a breakdown in the relationship between 'more' and 'better'. The only way to overcome these systemic problems is through a set of solutions which themselves address the whole.

In this report we have sketched out how, in the light of these challenges we face as a country and as a world, things could 'turn out right' by 2050. We have focused particularly on the UK, but many of the solutions we outline apply globally. We have called the process by which this could happen the Great Transition as a deliberate echo of *The Great Transformation*, written by Karl Polanyi in the 1940s.[1] While in a relatively short report such as this we could not hope to achieve anything remotely comparable to Polanyi's great work, the scale of the change we need to see is at least the equal of the changes he described.

Polanyi analysed how market processes in the industrial revolution had created severe ruptures in the fabric of social life, and argued strongly that we needed to reverse this and find a balance between the market and the non-market; the private and the public; the individual and the community.

We couldn't agree more, and the need to achieve this is all the more pressing now given the huge environmental problems we face, problems that Polanyi could not have foreseen in 1944 when the world was caught in the bloody conclusions of the Second World War. The most pressing problem facing humanity now is how to share scarce planetary resources in ways that are just, sustainable and support the well-being of us all.

What we have tried to do in this report is to set out how the balance Polanyi sought might be achieved in a modern context, particularly how this could be done in a way that safeguards the natural environment; where we live more in harmony with nature and within the 'carrying capacity' of our planetary resources.

We have named the key areas where the Transition has to happen. While the detail may change, we hope to show with this report that it is possible to envisage a just society where good lives do not have to cost the Earth, and to identify the levers of change that would need to be pulled to start the journey to this society.

A different future is not just necessary, it is also possible.

Stewart Wallis, Executive Director,
nef (the new economics foundation)

Executive summary

This report argues that nothing short of a *Great Transition* to a new economy is necessary and desirable, and also possible. Business as usual has failed. Yet prime ministers, finance ministers and governors of central banks are still running around – perhaps a little less frantically than they were – trying to allay fears and convince us that this is not the case.

The genie is out of the bottle though, and it is hard to see how it can be put back in.

The financial crisis exposed deep flaws in the approach to economics that has dominated policy-making for a generation. It turns out that letting markets rip does not always lead to the best outcomes for societies. Government intervention, far from being inherently inefficient, turned out to be essential to prevent system-wide collapse. A return to blind faith in markets to deliver a future of endless, rapid growth is impossible to imagine now.

This promise was always illusory, even on its own terms. Economic liberalisation has not delivered rapid growth on a consistent basis, but has been characterised by booms and crashes. Fewer and fewer have benefited from the booms as wealth and incomes have become increasingly concentrated, while those who suffered most from the crashes saw few if any benefits in the good times. Inequality has risen to record levels in many developed countries – not much has been 'trickling down'.

But the promise was also an illusion at a more basic level. Essentially, orthodox economic theory assumes the infinite consumption of finite resources. Growth, in the sense of using more and more stuff, is obviously impossible when there is only a certain amount of stuff. We can use what we have more efficiently of course, but this, too, has hard limits.

The world is warming. The atmosphere cannot absorb the levels of CO_2 being pumped into it for much longer without triggering irreversible climate change. The majority of the planet's ecosystems are being pushed to breaking point. Our 'footprint' in the developed world has grown too heavy, and we are showing no signs of our being able to tread more lightly.

Measures of life satisfaction in developed countries are flat. Overwork for many combines with widespread worklessness for others. Set alongside those who have far more than they need are those who do not have enough. Falling social mobility sees these patterns repeated from one generation to the next, while unsustainable levels of debt affect all parts of society. As real incomes have fallen, many have had to take o n debt to fund the essentials of life. For the more affluent, status-driven consumerism, often fuelled by debt, is the norm.

We can't afford to carry on as we are

Restoring business-as-usual – if such a thing is even possible – won't make us happy, and it will cost us dearly. This report forecasts that in the period to 2050 the cumulative cost associated with climate change will range from £1.6 and £2.6 trillion, while the cost of addressing social problems related to inequality will reach £4.5 trillion.

We appear to be locked into this state of affairs, but are we really?

It can still turn out right

We think that there is an alternative; one where we live within the limits of the natural world and more fairly with each other, locally, nationally and globally; where we focus on the things that really matter, applying our core human values to what is really valuable. We estimate that the measures proposed in this report would create up to £8.65 trillion of environmental and social value in the period to 2050.

This won't just happen. It requires us to rethink much of what we have taken for granted. As the well-off consume less, headline indicators such as GDP will have to fall by as much as a third, but we can grow 'real value' at the same time. By 2050, this increase in value would far exceed the fall in GDP, which is a very poor measure of 'progress' in any event. Rapid decarbonisation that moves toward global fair deal limits will avoid between £0.4 and £1.3 trillion in environmental costs. A progressive redistribution of incomes to reach Danish levels of equality will cut the costs of inequality-related social problems and increase social value by £7.35 trillion.

By sharing our resources more equally, by building better communities and a better society and by safeguarding the natural environment, we can focus on the things that really matter and achieve genuine and lasting progress with higher levels of well being. Taken together this would amount to what we have termed the *Great Transition*.

The Great Transition: in detail

In the first part of this report we make the case for change. In Part 2 we sketch out some of the steps we would need to take to make a reality of the *Great Transition*.

- In the *Great Revaluing*, we make the case that building social and environmental value should be the central goal of policy-making. We also argue that this needs to be true for private as well as for public decision-making, with market prices reflecting real social and environmental costs and benefits. We need to make 'good' things cheap and 'bad' things very expensive – too often this is the opposite of what we have today. As long as the achievement of good outcomes is separate from the real business of business, we will not see these outcomes achieved. Similarly, public policy cannot hope to create the best possible social and environmental outcomes unless this is at the heart of policy-making. In both cases, building real value requires us to accurately measure these outcomes and to build these measures into the core of public and private decision-making. This is a vital first step, upon which much else that is proposed in this report depends.

- In the *Great Redistribution*, we show how a redistribution of both income and wealth would create value as resources are moved from those who do not need them to those who do. We propose the creation of Citizens' Endowments of up to £25,000 for all people on reaching the age of 21 to enable them to invest in their future, as well as Community Endowments to provide commonly owned assets to invest in our local neighbourhoods. Both would be funded by a proposed increase in inheritance tax on all estates to 67%. As well as material factors, however, we also need to redistribute *time*. By sharing working hours and tasks more equally, everyone would be able to undertake more meaningful work, and by shortening the working week to four days we could create a better balance between paid work and the vital 'core economy' of family, friends and community life. We also propose a redistribution of *ownership* to create a form of 'economic democracy', where company shares are progressively transferred to employees in a resurgence of mutual and co-operative ownership forms. More equal societies are happier societies. By focusing on fairness we reap both social and economic benefits, as we no longer have to pay such a high price for the social 'ills' associated with high levels of inequality.

- In the *Great Rebalancing* we make a positive case for markets, but only once markets have been set up in such a way that prices reflect true social and environmental costs and benefits, and when those markets operate within scientifically defined limits. We also argue that the market sphere needs to be more tightly drawn and rebalanced alongside the public sphere and the 'core economy' – our ability to care, teach, learn, empathise, protest and the social networks these capacities create. In laying out the essential functions of the state, we again make a positive case – the state should be seen as 'us' and not 'them', and as a domain where we come together to achieve those things that are best done collectively. Arguing for a broader definition of 'public goods' and for the importance of maintaining low levels of inequality, we sketch out a facilitating state, which supports citizens, but also works with them to 'co-produce' well-being in areas such as health and education. This facilitating role requires a balance to be struck between direct provision, co-production, and the fostering of strong local relationships where people are encouraged to come together to pursue their shared goals and shape their own outcomes.

- This process is central to the ideas set out in the *Great Localisation*. Here we argue for an expanded concept of 'subsidiarity' – the idea that decisions are best taken at as local a scale as possible. This is enshrined in the principle, if not always the practice, of the European Union with regard to political participation and decision-making, which needs to be made more genuinely participatory and democratic but also more meaningful. By this we mean moving real power away from the centre to devolved democratic bodies and giving local people a real say in how this power is exercised. The principle of subsidiarity should also apply to the private sector. Redefining 'efficiency' beyond its narrow economic focus, we suggest a more rounded view, where the impact on the social fabric of cities, towns and rural areas is important when considering issues such as the production of goods and services. Exploring the question of what things are best produced locally, regionally, nationally and internationally, we suggest some criteria that might help in this judgement and make the case for greater local self-sufficiency in some areas, combined with regional, national and international trade in others. *Big* is clearly not always 'best' but neither, necessarily, is *small*. What we need is appropriate scale and, crucially, a clear means of deciding what this should be.

- The *Great Reskilling* continues this train of thought, starting from the position that greater local production will require us to relearn many skills that have been forgotten. From agriculture to manufacturing to the provision of local finance, returning to appropriate scale means equipping ourselves with the means to do so. Becoming less passive in terms of consumption and production we would start to regain our autonomy, which would extend to culture and arts, where we describe the beginning of a life-enhancing renaissance. This is not just the case for the economy and for the arts however; local decision-making based on active participation will be most effective when people are well informed about what makes their local economy tick and what makes public services able to achieve the best outcomes. Achieving consensus requires as full an understanding of these issues as possible.

- In the *Great Economic Irrigation* we outline how finance could facilitate many of the changes proposed in this report. For public finance we differentiate between tax and spending and between national and local levels. Developing criteria for what money should be raised at these two levels, we argue for a shift from taxing 'goods' such as work, to taxing environmental and social 'bads' such as pollution, consumption and short-term speculation. We argue for new variable consumption taxes, replacing income tax for the majority of the population, reflecting the social and environmental costs of goods. For private finance we again distinguish between national and local, arguing that large-scale projects such as building a green energy and transport infrastructure should be funded through national level environmental and 'land' taxes and the creation of public money where appropriate. This would be channelled through a national 'Green Investment Bank'. For private credit we suggest linking the ability of banks to create credit with the ability of borrowers to build social and environmental value, creating a 'race to the top' and reducing damaging credit bubbles. To get us out of the debt trap we now face, rather than slashing public services we propose a New National Housing Bank, offering people the opportunity to transfer a portion of their mortgage debt into equity and paying social rent on the balance. Locally, we argue for a restructured 'ecology of finance' of private, public and mutually owned institutions designed to meet local needs. In the local public sphere, we again suggest using the tax system to encourage social 'goods' and discourage 'bads' and argue that priorities should reflect local, democratically determined priorities.

- The *Great Interdependence* situates these national proposals firmly in an international context. While we focus on the UK in this report, we do this on the assumption of a particular global 'deal', which addresses global inequalities from both a development and an environmental perspective. Specifically, we assume that the essential global cap on carbon emissions to avoid irreversible climate change has been achieved and that the UK's share of this total carbon budget is broadly based on its population. As part of a phased transition we factor in a ten-year period where the UK reduces its emissions to align with the required 'convergence' path, during which time annual transfers to developing countries totalling around £200 billion are made. Mirrored by other developed countries as part of the same 'global deal', this total financing would add up to the trillions of pounds needed to enable developing countries to eliminate poverty and fund their own transition to a sustainable development path. Reducing total imports and reshaping the composition of these imports would obviously impact on national development strategies, but this process is already underway to some extent. The current global crisis has caused many in the developing world to question the wisdom of export-led growth focused on developed economies. While we are by no means suggesting that exports would cease, they would certainly be reduced as the environmental impact of transporting goods around the world was factored into prices. A rebalancing of internally and externally focused development is desirable for many reasons, and we would see more local production and more regional trade becoming the norm. Again, we are already seeing this begin to happen. If carefully managed and well funded, this would see development accelerate and 'stick', becoming more resilient in a stable and environmentally sustainable global context, with poverty and global inequality being progressively reduced and the huge dangers to developing countries of irreversible climate change averted.

These are big assumptions, but we do not apologise for that. A global deal along the lines outlined here is essential for environmental reasons, but also to finally rid the world of the scourge of poverty and inequality. Business as usual has also failed in this regard. Just as within countries, trickle down approaches at global level have brought us to the brink of environmental disaster, while also increasing inequalities and entrenching grinding poverty in many parts of the world. Locally, nationally and globally we need to change direction quickly and radically. We need nothing short of a Great Transition – to collectively build a different future.

This report concludes with a discussion of two big challenges that have to be addressed before this can be achieved, but also sets out clear steps that can be taken straight away to start the journey. While there is much that the Government needs to do, there are also things that we can all do now. We cannot afford to wait and neither should we. The possible future sketched out in this document is not intended to be prescriptive in any way, but to show that not only is fundamental change possible, it is also very appealing. We might have to give some things up, but these are not the important things in life. What we could gain, on the other hand, would be something really worth having.

The *Great Transition* can start here.

A tale of how it turned out right

A tale of how things turned out right. There we were back in 2009 feeling as if the four horsemen of the apocalypse were bearing down on us. The global economy was falling apart, we were accelerating towards the cliff edge of catastrophic climate change, and our oil-addicted economies were set to go cold-turkey as their fossil-fuel fix grew much more expensive and harder to get hold of. At the same time the world was divided by great wealth and extreme poverty; overwork and unemployment; hunger and obesity; and even the relatively rich global minority found themselves consuming ever more but without any noticeable rise in their well-being.

It was all going horribly wrong. Then, suddenly, common sense kicked in. We realised that we had one chance left. If we blew that, there could be no turning back, no excuses. It wasn't as if we hadn't been warned, or that we didn't know what was going wrong. Change became the name of the moment. The Great Transition began. A future was mapped out that would allow the country to live within its fair share of global environmental resources. We even worked out how to manage the Transition in a way that improved our well-being, enhanced community life, and promoted social justice.

And, after it all turned out right, this is how an average, working urban-dwellers day in the UK panned out.

With less time spent working in the formal economy and more flexibility over when we work, the choice is ours – take the kids to school, go for a run, read a book. With a new focus on real wealth and well-being, previously overconsuming rich countries have now cured most cases of the twin evils of work addiction and unemployment. The huge debts and interest payments that kept us chained to our desks have been designed out of the system by new forms of credit and ownership, for land, homes and other big ticket items. Because we're more content, having more time for ourselves, friends and family, we need less income too for the false consumerist promise of buying happiness. More flexible working practices have made it much easier for us to work part-time, take sabbaticals and tailor where and when we work. We're using technology cleverly to make for smart work. Those of us choosing the early morning run enjoy fresh air in our lungs and clear paths as dramatic reductions in traffic have transformed city air and streets – the result of a successful shift to mass transit systems and the new popularity of walking and cycling.

When we do need to buy things, there's no call to sweat over every shopping decision: business and Government have got it together to make socially and environmentally sustainable trade the (carefully checked) norm. They have also ensured that both work and incomes are much more equally shared. The weekly food bill has gone up – but so has the quality, and so has our ability to pay for good, healthy food. Also we're saving lots of money later in the day, and are no longer struggling under unsustainable levels of personal debt. The damaging consequences of cheap food systems have gradually been rolled back. This is sustainable consumption universalised – no more scanning labels. A few deft and well-planned moves in boardrooms, Parliamentary chambers and the Inland Revenue helped to make food markets fair and sustainable.

Flexibility and technology have massively reduced our need to travel for work. The hours gained and stress lines postponed make us more effective and committed to the work we do, particularly as our successful organisation is likely to be one creating real social and environmental value. But these changes are about more than work. Social networking software has thrown us together with new people– our desktops give us a global network, but also connects us in new – live – human ways to the communities in which we live.

For those of us happy to live without a computer, there are plenty of benefits in the new sense of community that has evolved from the revival of local shops (where the shopkeepers actually remember who we are) and the way that residential streets and town centres, liberated from suffocating traffic, have become people-friendly. Streets are safer for children to play in, with some entirely car-free, and many towns have reclaimed central plots of land as public squares. A calmer environment and more opportunities for casual contact between neighbours make public space more accessible to all. People of all ages gather and talk to each other more and, as a result, even in cities people, particularly older people, feel less lonely and vulnerable. Crime has fallen too.

We can take some time out late morning to plan our summer trip. While the big increase in the cost of fossil fuels has seen international travel become a much rarer experience, it tends to be much better – and longer – when we do head off on our travels. With more leisure time and good cycle and public transport links, low-impact local excursions are a much-loved part of many people's lives. But with our experience of both cities and countryside transformed by investment in really great public spaces – whether it's the park or local recreation ground, the village hall, local pub or café, theatre or cinema – we feel less need to get away in order to unwind.

A journey to work? Problems are as big as we make them: it used to be said that we wouldn't give up our cars. Cars were bought; roads were built; resources (including our own wallets) were burnt in pursuit of a very particular form of mobility that becomes less enjoyable and more polluting the more people take it up.

But by raising revenue from polluting and inefficient fossil-fuel-run cars to invest in alternatives, governments were able to completely transform people's experience of cities and towns. Owning and driving cars to meet most of our mobility needs has come to seem simply eccentric. Lifespan and quality of life have dramatically increased. Transport options range from trains, trams and quiet clean buses, to on-demand rural shared taxis and simple car-share schemes that meet the range of needs we have throughout a year.

In the evening, time released from long working days, and the fact that fast food and ready meals have gone up in price now that they reflect their full ecological costs, has seen a revival of home cooking. With lots more single households there are some twists. More people get together to take turns to share informal meals in a neighbourhood. In fact, everywhere, people are relearning skills that for much of human history were second nature, but which had been largely lost in just a couple of generations.

Stories and music are as old as campfires. For a time we forgot it, but being actively involved in making entertainment made us feel much better than just passively watching others perform. Perversely, though, the fashion for reality TV talent shows early in the twenty-first century triggered a widespread revival in people wanting to do things for themselves so, in any case, we started to spend fewer and fewer hours in front of the television. It's now common in pubs, clubs and in any available hall to find groups of friends showing films they have made on inexpensive, easy-to-use equipment, and putting on a wide range of music and other performances.

The urge to take part goes beyond arts and culture. Even politics has become a way for us to come together and make our voices heard. Political power, once locked away in distant centralised institutions, is now embedded in local communities. Taking part in politics now means debating issues with our friends and neighbours; open public meetings where anyone of us can have our say on the issues that matter to us.

Just as people are happier to go out more locally during the day, because towns have become more pleasant places to be, the same is true at night. As in countries like Italy, in the early evening people of all ages take to strolling around town, just for the sake of it. The increase in spare time means people start reviving half-forgotten festivals and celebrations, as well as creating new ones to mark everything from important global events, to the seasons, local history, people and important events. There is much more partying in general, and not just for the young.

A revival of distinctive local economies also brings more character back to different areas, making it worth travelling around the local area to visit other unique local festivals, bars, restaurants, cinemas and theatres. Clone towns dominated by identical chain stores and outlets are consigned to history.

Part 1:
Why we need a Great Transition

Rediscovering the purpose of it all

It is preoccupation with possessions, more than anything else, that prevents us from living freely and nobly.

Henry David Thoreau, *author* (1817–1862)

It is a virtual axiom of orthodox economics that *more* is always *better*. At national level, all considerations pale beside the need to keep GDP figures ticking upwards. At the individual level, the pursuit of individual wealth has risen beyond mere social acceptability to become synonymous with 'aspiration' and even, in some quarters, touted as a moral good.[2]

Yet, even the briefest trawl through the history books suggests that this is profoundly at odds with centuries of philosophical, moral and religious teaching on how to live a good life. Throughout history, across widely different cultures, people have recognised the dangers of an excessive focus on wealth and material possessions. Viewed in this wider historical context, our recent fixation with the pursuit of individual gain as *the* route to happiness looks a curious anomaly.

It is curious chiefly because, just as the sages of history told us and recent empirical research repeatedly confirms, once people's needs are met, their material circumstances play a relatively small role in determining their well-being. Of course, given that half of the world's population lives on less than $2.50 a day and one in thirteen children die before the age of five, getting everyone to this point is not a trivial matter. There can be no doubt whatsoever that, in many countries, raising material standards of living is an urgent requirement.

But once this point has been reached – as it could be already for many in the developed world if our resources were more equally shared – wealth and possessions simply don't make much difference. People *adapt* quickly to changes in material circumstances. When you receive a pay rise, move house, buy a new television set or the latest must-have fashion item, you'll probably feel happier for a while (although, in fact, probably not as much or for as long as you expected[3]). All too soon the novelty wears off and with it any lasting benefit. This is not to say that having these things is not desirable, or that acquiring them is not pleasant, but simply that they make very little fundamental difference to the things that really matter to our lasting happiness. This 'hedonic treadmill' is responsible for the rising expectations that come with wealth but also for the persistent dissatisfaction of never seeming to have enough.

Much more significant to our long-term happiness are factors relating to individual differences in outlook and to the kinds of activities that people engage in: having and raising children, socialising, participating in cultural life, caring for relatives or friends, and having a sense of meaning and purpose. These things have little to do with material possessions (provided, again, that basic needs are met) and everything to do with social and cultural norms and values. In fact, it is well established that having an overly materialistic outlook on life has negative impacts on well-being. Materialistic people suffer worse outcomes across a range of domains: happiness and satisfaction, mental health, social relationships, and more.[4]

At the heart of this report is a fundamental belief: that the 'purpose' of the economic system is to improve well-being for all within the limits of what the planet can sustain – to produce good lives that don't cost the Earth. When we think about the kinds of changes that will be required to bring this about, it is tempting to focus just on the practical issues – financial regulation, taxation and welfare policy, or reducing our carbon intensity – and we will be coming to these issues in detail later. But we need to remember that, as important as these are, in a democracy, none of these changes will come about without the will and desire of the people. And people are not like the passive automatons of economics textbooks. They have goals, beliefs and aspirations and they actively construct the world around them through the ways in which they talk, behave and make meaning.

Whilst it has become unfashionable to talk about values in arguments for socio-economic change, we cannot neglect this issue.

This is not to imply that our current values are somehow 'broken'; rather, it is a matter of rediscovering what has been there all along. The aspirations compatible with a healthy planet and society are already part of most of us. Are we shopping because we think this, in and of itself, will make us happy? Or is it for other reasons, perhaps because we believe that having nice things to wear is important to our success in social networks? If the reason is something like the latter then it is friendship, not shopping, that we really value.

Rediscovering, in this way, what it is that matters most to us creates an opening for thinking about how we might achieve those same ends but without the huge attendant costs for the planet and society.

Sketches of the future such as the one presented earlier illustrate one – and *just* one – way in which the values we share might be expressed differently. In our sketch, people value their time more than their income, and so institutions and working practices make it possible for them to choose to spend more time with their families, in their communities or pursuing their interests. Occupational opportunities are shared much more widely so that there are few people who can't find paid work. Instead of paid employment being the predominant way in which people feel able to demonstrate their value, in future people find reward and pleasure in doing things for themselves and value sharing their skills and assets, so community is enhanced and the market remains in those areas where it serves a productive purpose.

If the practical details of this vision seem a long way from where we are now, it is worth noticing that the values and aspirations it reflects are not. We've always known that more isn't always better.

The big challenges: why we need a Great Transition

The events of the last year have made clear the scale of the challenges we face – we really do need to get this right and start doing so soon. Most visibly, the banking meltdown triggered the most severe economic crisis since the Great Depression. But this is by no means the only one. Greenhouse gas concentrations are reaching levels where runaway climate change becomes more and more difficult to avoid. Peak oil may be far closer than we think and there is little preparedness for the energy crisis that lurks just around the corner.

Inequality between countries, as well as within countries, is reaching new highs. On top of all this, life satisfaction even in the most developed countries is at best stagnant, and even declining in some. It is hard to conclude anything other than that the current model of pursuing economic growth at all costs – environmental and social – just isn't working.

The clock is ticking: climate change, peak oil and ecosystem pressure

There is an overwhelming scientific consensus that the threat posed by climate change is serious and urgent. The Earth's climate system is changing at greater rates and in patterns that are beyond natural variations. We are witnessing changes to the temperatures of the atmosphere, land and ocean, the extent of the sea-ice and mountain glaciers, sea level, the distribution of rainfall, and the length of the growing seasons. All of these, at least in part, have been attributed to the rapid increase in the abundance of man-made (anthropogenic) greenhouse gases.[5]

In addition to the physical impacts of climate change, over the past 150 years humans have had a huge impact on the Earth's biosphere, largely as a result of rapidly growing demands for food, water, timber, fibre and fuel. The damage to ecosystems has now resulted in a largely irreversible loss in the diversity of life on Earth. These 'home-made' extinctions, driven by human activity are the most extreme they have been in the past 65 million years.[6] The species we are losing are part of the ecosystems which provide for human livelihoods and welfare: everything from the food we eat, to the air we breathe, the medicines that keep us alive and the landscapes we gaze upon to find solace and inspiration.

Strongly coupled to climate change and overlapping in terms of its critical timeframe is the 'peaking' in global production of oil and gas. The so-called 'Hubbert Peak' theory, also known as peak oil, or the topping point, describes a point in the path of the extraction and depletion of conventional oil and other fossil fuels at which world oil production will soon reach a peak, level off and then rapidly decline.[7]

Climate change

In 2007, the Intergovernmental Panel on Climate Change (IPCC) Fourth Assessment Report – a synthesis of peer-reviewed research on climate change, its causes and effects (including socio-economic consequences) involving over 2500 scientists worldwide – stated that if fossil fuels continued to be burnt at the current rate, global average surface temperatures could rise by 4°C by the end of the century, with an uncertainty range of 2.4–6.4°C.[8] While this implies emissions need to be reduced dramatically and immediately, there is growing evidence that even a rise of 2°C – the widely accepted *maximum* 'safe' level – could be catastrophic for ecosystems, humans and economies.

The 2°C target has been widely accepted as the guiding principle for many climate change policies – for example, it has been the EU negotiating position at climate change talks since 1996. However, Professor Stefan Rahmstorf from the Potsdam Institute in Germany argues: 'If we look at all of the impacts, we'll probably decide that two degrees is a compromise number, but it's probably the best we can hope for.'[9]

Recent research by climate scientist James Hansen supports this view. Hansen and his team find that that temperatures should not go beyond 1.7°C (or 1°C above 2000 temperatures) if we are to avoid practically irreversible ice sheet and species loss.[10] But, in terms of the social impacts of climate change, what is manageable for some is actually catastrophic for others. For example, small island states argue that 1.5°C is a better target, as many of them will disappear with warming beyond this point.[11]

And time is running out. In 2008, cautious calculations by **nef**'s climate change and energy programme suggested that there may be as little as 100 months, from August 2008, to stabilise concentrations of greenhouse gases in the atmosphere before the risk of uncontrollable global warming occurring increases significantly.[12] **nef**'s analysis has been supported by the recent research of the Tyndall Centre for Climate Change Research, and is supported by the chair of the IPCC, Rajendra Pachauri.[13]

Worryingly, the latest figures from the Global Carbon Project, an international collaboration between leading climate research institutions, shows that emissions are fast heading in the wrong direction. They estimate that the average annual growth rate in emissions was 3.5 per cent between 2000 and 2007.[14] At the same time, globally, there is no sign of a slowing in the growth of emissions. There has been a constant or slightly increasing trend in the carbon intensity of energy (carbon emissions per unit of energy) over recent years, in both developed and developing nations.

This means that each time governments congratulate themselves for achieving 'record levels of economic growth', global atmospheric concentrations of CO_2 conspicuously lurch forward.

Peak oil

The threat of climate change should be a sufficient cause to tackle our carbon-intensive lifestyles, starting with our dependence on fossil fuels. There is also another very real and pressing reason to do so: peak oil. Opinion is divided as to when peak oil will be reached, but some experts believe it could be imminent, or may even already have passed.

Energy prices reached record highs in mid-2008, culminating in a 14-fold rise in the price of oil from 2000 levels. By July 2008, oil prices had reached $147 a barrel. While prices fell in late 2008 to $40, only to rise yet again to over $70 in 2009, there is a growing consensus that oil prices are becoming increasingly volatile due to declining domestic reserves and production and growing reliance on international markets that are concentrated in a relatively small number of regions. For example, it is estimated that 62 per cent of all known oil reserves reside in the Persian Gulf.[15]

The International Energy Agency (IEA), hardly a radical body, first acknowledged peak oil in its 2006 *World Energy Outlook*. In 2008, the IEA went further, stating that there would most likely be a 'narrowing of spare capacity to minimal levels by 2013'.[16] While this might at first sight seem welcome to those concerned about climate change, the social and economic consequences of being forced to go 'cold turkey', rather than managing the transition from dependency on oil, could spell disaster. Until there is investment in viable alternative sources of energy, and our lifestyles are able to adapt, we are all locked into oil dependency: as we showed in *Nine meals from anarchy*, even our food supply depends on it.[17]

Despite huge untapped renewable energy sources in the UK and Ireland, both have become increasingly reliant on imports and struggle to maintain their energy independence. The UK is increasingly vulnerable to wild fluctuations in oil and gas prices. This will have huge impacts on social justice. For example, the Government estimated that for every 1 per cent increase in the price of energy, a further 40,000 households will be plunged into fuel poverty.[18]

Ecosystem pressure

Our environmental challenges extend well beyond carbon however; Earth's life support processes depend on the optimal functioning of ecosystems. Human life is dependent on the regular availability of food, water, shelter, optimal atmospheric conditions and nutrient recycling systems. It is this interaction of biological, physical and chemical elements that guarantees the provision of everything humankind needs to thrive on earth. The natural world is also key to mental and physical well-being – a large body of research from Europe and North America shows that people and in particular children derive significant psychological benefits through exposure to nature and that exercise in a natural environment is more beneficial than in urban environments.[19,20]

Yet forests, food supplies, water, marine life and many other natural resources are under threat from over-consumption by those in developed countries (Box 1). For everyone to live at the current European average level of consumption, we would need more than double the biocapacity actually available – the equivalent of 2.1 planet Earths – to sustain us. If everyone consumed at the US rate, we would require nearly five. Neither of these is a viable option: consumption in the developed world must be cut back to preserve the ecosystem and enable growth of living standards in the developing world.

Of course, the pressures we place on environmental resources through consumption often have what scientists term – without intentional irony – 'positive feedback' loops. Consider soya. In 2005, the UK imported 774,623 tonnes of soya into the UK, with around two-thirds coming from Brazil.[21] Much of it goes into animal feed to support our relatively new-found habit of having plenty of meat and dairy products in our daily diet. But soya production is one of the key pressures that, along with forest fires, drought, deforestation and climate change, could push the Amazon rainforest over a tipping point where, rather than being a store of carbon dioxide, it begins to release it. With the Amazon estimated to store 120 billion tonnes (± 30 billion) in biomass carbon,[24] scientists have said this switch could trigger 'runaway climate change'.[25,26] If the environment is at breaking point, and with time running out, what about the economy?

Box 1.
Natural ecosystems loss and human well-being[22]

The Millennium Ecosystem Assessment led by the United Nations Environment Programme (UNEP) found that 60 per cent of a group of 24 ecosystem services they examined are being degraded or exploited beyond ecological limits; they also found evidence that these changes increase the chances of non-linear changes in ecosystems. Examples of these include the collapse of fisheries, the spread of disease, loss of water quality, and shifts in regional climate.

Key findings are:

- More land was converted to cropland in the 30 years after 1950 than in the 150 years between 1700 and 1850.
- The world has lost 50 per cent of its wetlands since 1900.
- The world has lost 35 per cent of mangrove area since 1980.
- 20 per cent of original area of coral reefs has been effectively lost since 1950 and more than 20 per cent are badly degraded or under imminent risk of collapse.[23]
- There is four times more water in dams now than in 1960 and three to six times as much water is held in reservoirs as in natural rivers.
- Agriculture accounts for 70 per cent of worldwide water use.
- The current species extinction rate is about 1000 times over background rates over the planet's history.
- Forest area has shrunk by about 40 per cent over the past 300 years. Twenty-five countries have completely lost their forests and 29 countries have less than 10 per cent of their forest cover.

An economic model that failed on its own terms, and for people and the planet

The neoliberal economic model has failed spectacularly. What started as a run on banks and saw major financial institutions, most symbolically Lehmann Brothers, collapse, turned into the most severe global recession since the 1930s. If the theories of self-correcting and efficient markets had been right, the events of the last 18 months could never have occurred. But they clearly did. What we have seen is not just a temporary malfunctioning of the neoliberal model but its failure on its own terms. Instead of endless, stable growth and high and rising incomes equitably shared, we have had inequity, volatility and crises. These are not anomalies, but a natural and increasingly severe expression of the 'normal' functioning of the system. As even Alan Greenspan, former Chair of the US Federal Bank, was forced to admit, there was 'a flaw … in the model that defines how the world works'.[27]

But the economic model has not just failed on its own terms. It has also failed people and the planet.

The environmental and economic crises are not separate but interconnected events. It is the high levels of debt-fuelled consumption in developed countries that have landed us with dangerously high concentrations of CO_2 and put pressure on ecosystem resources. Astonishingly, this is precisely the path that politicians are trying to return us to. Many of the measures hastily put in place at the start of the recession – VAT reductions and the car scrappage scheme, for example – were specifically designed to kick-start consumption.

While there is a clear case for poor countries growing their economies to improve living standards, increasing consumption further in developed countries is not sustainable or justifiable. In fact, developed countries need to do the exact opposite.

The fact that countries like the UK need to cut back, rather than increase, consumption should not worry us from a well-being perspective. After all, as we saw in the last chapter, the path of ever-increasing consumption is not making us happier. If anything, the pursuit of consumption leads to lower levels of well-being.[28]

Beyond the evidence on life satisfaction and the environment, there is another reason why the attempt to resurrect the old economic model is a bad idea. Far from the mythical promises of trickle-down, the growth model has seen huge inequalities develop within and between countries. In 2005, more than half (56.6 per cent) of the world's population lived on less than equivalent of $2.50 a day. Research conducted by **nef** showed just how unevenly the proceeds of growth are shared out. For every $100 worth of growth, only $0.60 contributes to reducing poverty for the more than one billion people living on less than a $1 a day.

As stark as these inequalities are on a global scale, we should not ignore inequality within countries. And the UK has one of the highest rates in the developed world. The richest 20 per cent have seven times the income of the poorest 20 per cent. One in three children grows up in relative poverty. All this makes for a profoundly unequal society, which matters from a social justice perspective but also because, as Richard Wilkinson and Kate Pickett demonstrate in *The Spirit Level*, less equal societies have poorer outcomes in nearly every social domain.[29] And this holds true for people across the income spectrum. So while those on low incomes obviously have a disproportionate share of the poor outcomes, a middle-class person living in a country with high inequality will, for example, have a lower life expectancy than someone of the same socio-economic status in a more equal society.

So what if business as usual continued to 2050?

It is hard to see how it could be done but, for arguments sake, what might the UK look like in 2050 if politicians did manage to restore business as usual?

Figure 1 projects what could happen to GDP and to costs associated with climate change and inequality. We have been optimistic in our forecasts – this is not an attempt to deliberately construct a frightening scenario, but to give a clear-eyed picture of a possible future. GDP has been plotted as following a steady growth path and we assume that inequality remains at 2006 levels. With the climate change costs, however, we have modelled two separate cases. One – scenario B – takes politicians at their word and plots what could happen if the Climate Change Act targets of an 80 per cent reduction by 2050 and maximum concentration of 550 parts per million (ppm) are reached.

Given the rate at which emissions are increasing and the lack of concerted policy measures being put in place to try to reverse this, we felt that it necessary to also project a more realistic scenario – scenario A – and have used the projections of the World Energy Outlook (WEO).[30]

Even in the 'optimistic' scenario where Climate Change Act targets are met, the UK is by no means a happy place by 2050. While GDP is certainly higher, this has been bought at a quite phenomenal price. The cumulative cost of climate change and the cost of abatement have reached £1.6 trillion, which gets much worse when we consider the more realistic WEO scenario. Here the cumulative costs by 2050 are expected to reach £2.6 trillion.

Also, by failing to address the high levels of inequality in the UK, we allow the associated social ills, such as poor health outcomes and high levels of crime and family breakdown, to continue at the high levels seen today. We have plotted the annual costs of this in Figure 1 because the cumulative total would dwarf the other amounts and distort the graph. The cumulative total over this period for the cost of social ills associated with inequality exceeds £4.5 trillion.

There must be a better way than this.

An alternative scenario for 2050: from crisis comes opportunity, creating real value

What if we begin not from growth as our goal, but with a goal of maximising real social value for all within the confines of what is possible, given the planet's carrying capacity and the UK's equitable share within these limits? In order to do this we would need to produce and consume less, probably quite a bit less. GDP as it is calculated now will fall, maybe by as much as a third in our estimates, which only takes us to the level of GDP we had in 2001.

People rightly associate falling GDP with rising unemployment, poverty and general misery. The way our economy has been structured is such that this has indeed been the case. A central goal of the Great Transition is to manage this change in such a way that we are better off not worse off. We think that this can be done. GDP is a very poor measure of progress: the revenues skimmed off the financial system by traders in the City of London as the pyramid of 'toxic' derivatives was being built added to GDP; cleaning up the effects of pollution increases GDP; paying the costs of high rates of crime increases GDP. None of these things can be said to build lasting social or environmental value. Rather, they are highly destructive of it.

By focusing on building real value within environmental limits we believe that life can be better for individuals and their families; that society can be a better place to live for all; that our natural environment can be protected for both current and future generations. It will not be easy, but it can be done, even if GDP falls some way so that we use only our fair share of sustainable global environmental resources.

Figure 1: Business as usual scenario depicting GDP, cumulative costs associated with climate change and annual cost of social ills associated with inequality to 2050.

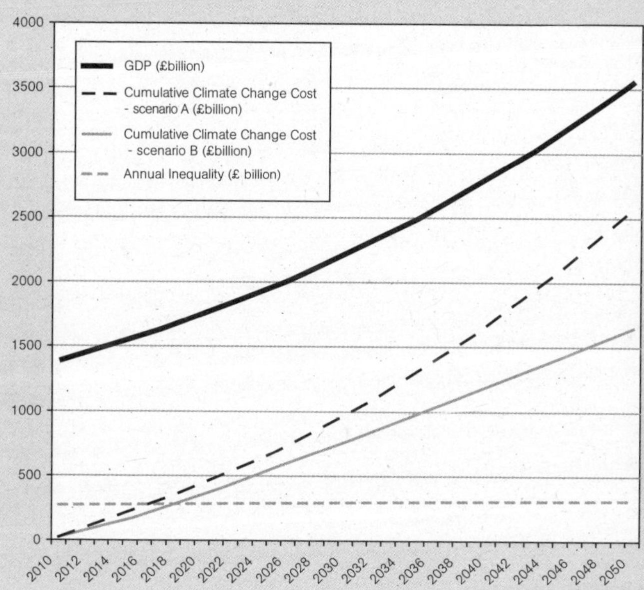

As well as a shift in mind-sets and a return to the core values that define us, this will require us to think much more intelligently about how we organise our economic and social affairs than we have done till now. We need to make much better use of the resources – social, environmental and economic – that we already have. It requires a Great Transition from where we are now to where we need to be.

Realising the Great Transition that is sketched out in this report would see value rise steadily across three dimensions: for individuals, through a more equal distribution of incomes and assets; for society, by reducing the costs associated with inequality; for future generations, by protecting the natural environment and so reducing climate change-related costs.

First, though, a health warning. This is an imperfect method of valuation, as all assessments of value are subjective and therefore fraught with difficulty – how much a pint of milk is worth to me will vary depending on my circumstances. This does not make it a futile exercise; it is an illustrative example to demonstrate how the pursuit of well-being and wider social and environmental value could be achieved without growing GDP. In the absence of anything better, monetary valuations are used as an approximation – a 'vaguely right, rather than precisely wrong' approach – not to place a fixed price on something but using money simply as a common unit of account in our calculations. This is, of course, a tool from orthodox economics but by adapting it to measure value that is usually hidden and so ignored, we disrupt the relationship between money and value. For example, for people on low incomes we recognise that each extra £1 is worth more to them than to people on high incomes, and adjust our valuation accordingly.

The valuation exercise outlined below is highly instructive because it reveals that, for as long as we fail to account for hidden forms of value creation, we will continue to rush headlong towards destroying the very conditions that make civilisation possible.

Figure 2 shows how value progressively increases across all three domains from 2010 to 2050. So how is this value created?

Individual value

Even from an orthodox, neoclassical economics perspective, unequal societies are highly inefficient in terms of utility maximisation. Utility is a central concept in economics and refers to the satisfaction that is derived from the consumption of goods and services. It is odd, given the importance of utility in economics, that economists have paid very little attention to the total utility loss that occurs when income is distributed unevenly. Yet it is a simple concept. Think about it. An extra £500 holds little additional value for someone in a household where the income is already £100,000. But it makes a big difference to someone in a household where the income is £9,500.

An important part of the Great Transition is a progressive redistribution of incomes. Income is transferred from high-income to low-income households over a period of 20 years in five waves. This is done in such a way that post-redistribution, no household has an income of less than £13,900 – a figure that a recent study by the Joseph Rowntree Foundation identified as the basic minimum for the UK.[31] The resulting income distribution is far from unrealistic. It is actually the same as Denmark today.

Figure 2. Cumulative value creation in three dimensions: individual, social and environmental.

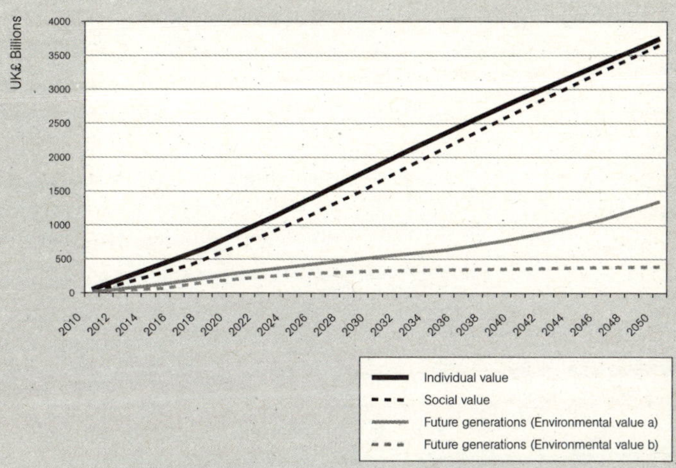

By moving money from where it is not needed to where it is, we increase overall utility in the economy, increasing total value by giving more income to those individuals who need and value it most.

Our model calculates the additional value that is created by moving money from households where little utility is derived to households that will get greater utility – i.e., to where it is actually needed. The percentage change in moving from pre- to post-redistribution is 8.49 per cent. This is the equivalent to a cumulative gain of £3,752 billion from 2010 to 2050, as shown by the solid black line in Figure 2. (See Appendix 2 for details.)

Social value

Unequal societies are not just bad for people at the bottom end of the income distribution – they are bad for everyone in society. Unequal societies have worse outcomes across nearly every social domain: health, life expectancy, education, and crime, to name but a few.

These outcomes are undesirable in and of themselves – few would choose to live in a society with very bad outcomes in these areas if they could do something about it. But they are also a significant and preventable expense to the public purse and to the economy as a whole. Given the established link between levels of inequality and the incidence of these outcomes,[32] we believe that we can build real social value by reducing inequality and so reducing the prevalence of associated social ills.

By building a more equal society through the Great Redistribution, the incidence of these poor outcomes is reduced. Our model places conservative estimates on the social value of greater equality from reduced social problems in areas such as health, education and crime. This value includes not just the savings to society from reduced dependence on public services to deal with the impacts of these problems but also the benefits to people and communities from not having to live with the consequences of high crime, low attainment, family breakdown and ill health.

Because our post-redistribution inequality is close to that in Denmark, we have used this as the benchmark – i.e., we have assumed that outcomes converge on those experienced in Denmark, controlling for population size and calculating the resulting savings in UK terms.

Over the 40 years to 2050, we estimate that this would create cumulative social value of £3,655 billion as shown by the pink line in Figure 2. (See Appendix 3 for details.)

Future generations (environmental) value

The final form of value creation is environmental. Put another way, it is the value that would be preserved for future generations in the Great Transition relative to two possible business as usual scenarios. In each case we estimate the environmental impacts of business as usual (scenario A and scenario B) and the Great Transition, with the difference between the two being the environmental value created (or degradation avoided).

In order to avoid climate-change-related costs spiralling out of control, reaching a cumulative total of £2.6 trillion by 2050, the Great Transition sets out plans for a rapid decarbonisation of the economy that keeps global atmospheric carbon dioxide concentrations to 450ppm. While this figure is criticised in many political circles for being unrealistically ambitious, many scientists and progressive governments no longer consider this adequate to prevent dangerous climate change.[33] 450ppm is used here because it derives from an available model and still reveals the characteristics of rapid (if not deep enough) decarbonisation. In this scenario we assume maximum possible efficiency in terms of energy use across all sectors of the economy, and lowest possible carbon intensity of energy as currently understood.

Moreover, we assume that the UK progressively reduces its carbon footprint so that it uses only its fair share of total global carbon emissions under the given, interim target, making sure that other countries, particularly developing countries, have space to develop and make their own transition to a sustainable future. We assume a global 'deal' based on 'contraction and convergence'[34] to limit, reduce and maintain total global emissions within defined limits (the contraction); we also assume that the UK's total share of emissions progressively comes into line with its fair global share (the 'convergence'), with significant transfer payments to developing countries during the process to facilitate their sustainable development.

The business-as-usual scenarios, in contrast, are based on emission reductions in line with an atmospheric carbon dioxide concentration that peaks at 550ppm. But most scientists would say that we stand little chance of preventing runaway climate change at such concentrations, so that our value is created by avoiding the costs associated with the higher concentration level.

The business-as-usual scenario B assumes that the Government achieves the targets set out in the Climate Change Act. The environmental value created (or the lower environmental costs accrued) relative to this scenario if the Great Transition is implemented totals £413 billion from 2010 to 2050.

Our less optimistic business-as-usual scenario A is one where the Government fails to achieve the targets laid down in the Act but instead recent trends continue broadly as now. This accords with projections made in the WEO described earlier. Given the lack of progress to date this is a far from unrealistic scenario. Relative to this, the cumulative environmental value created (or the difference between the costs accrued under this scenario and the Great Transition) would be £1,345 billion over the same period. (See Appendix 1 for details.)

Total value vs GDP

We have shown how we can build individual, social and environmental value in the Great Transition. To achieve the latter and to avoid catastrophic and irreversible climate change, which is a prerequisite for building any form of positive society, we have to live within our environmental means as a country. This means that we need to produce and consume fewer resources, which will result in a reduction in GDP as it is measured today.

There really is no alternative, but we can do this in ways that actually increase real value for individuals, for society and for future generations, too. As shown in Figure 3, the total value created in each of these three domains exceeds the cumulative reduction in GDP of £7,360 billion from 2010 to 2050. This equates with a phased reduction in annual GDP of about a quarter, from £1,384 billion in 2010 to £1,025 in 2050.

If the Government sticks to the commitments of the Climate Change Act, total value created through the Great Transition would be £7,820 billion. If current trends are maintained, the figure rises to £8,752 billion as a result of the larger environmental value created (or destruction avoided) under the Great Transition.

In order to compare this value with GDP we have expressed it in monetary terms. It is important to stress, however, as above, that in each case – individual, social and environmental – the monetary values are 'proxies' for real underlying value. Taking money from people who don't need it and giving it to those that do will enable them to buy things they need for themselves and for their families. Our process of redistribution has ensured that no one receives an income below that identified by the Joseph Rowntree Foundation as the minimum basic income required to buy the goods and services deemed essential. Thus we take from those that have too much and give it to those who cannot afford the basic essentials of life. The 'value' shown in Figure 2 is therefore a monetary representation of the magnitude of this real increase in individual value aggregated across society.

Similarly, social value is a proxy for the improvements in people's lives that would result from better outcomes in health and education; from lower crime and safer streets; and from more stable family structures. These are things that matter to people; things which go a long way in determining the sort of society we all would want to live in. Some of this is a direct cash savings to the state – e.g., lower costs associated with preventable health or crime – and some is an estimate of the social value these improvements would make for citizens in their neighbourhoods. The monetary estimate is therefore a proxy for these real increases in value, which reflects their potential magnitude and presents this in a common unit of account – in this case, UK £s.

Environmental value is also a proxy, which represents some of the value preserved by mitigating the effects of climate change and hopefully (the precise sensitivity of

Figure 3. Total value creation vs GDP reduction 2010–2050.

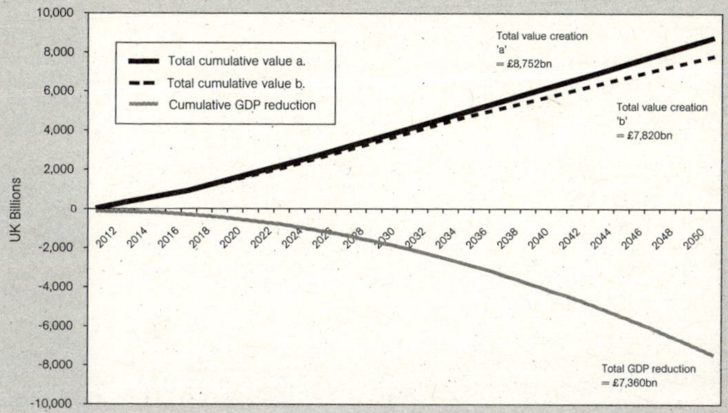

ecosystems is still highly debated) avoiding it becoming catastrophic and irreversible. This value would accrue to current generations by preserving the natural environment in a manner where we can continue to use its vital resources to sustain ourselves, but also to future generations as we would do so in a way that is compatible with the planet's 'carrying capacity' so that they can enjoy these same benefits.

For both current and future generations, however, it is not just using the Earth's resources that will create value, but simply being able to live within and experience the wonders of nature in healthy and sustainable ecosystems. We do not pretend to have accurately captured these different forms of value – or to suggest that such a thing is ever exactly possible, but we do know that if we degrade the environment we destroy value. Conversely, if we safeguard and preserve the environment, we create lasting environmental value for current and future generations. Our monetary estimates reflect estimates of the real economic benefits (or absence of costs) to our societies of doing this, and so are a rough proxy for the magnitude of real environmental value created, again simply expressed in a common unit of account.

And making us happier too…

Growing value across these dimensions is essential for providing people with the fundamental prerequisites for living a good life and for creating the conditions for all individuals to flourish.

By creating safer and better-functioning neighbourhoods we create the conditions for meaningful personal relationships to be forged. By improving health we help to enhance overall well-being by boosting people's vitality. By addressing levels of inequality in society we target issues of 'relative status' which can undermine people's sense of competence and meaning in life. And by reducing consumption we may reduce the prevalence of materialism which is so strongly associated with negative well-being outcomes, and create space for other, more fundamental values to come to the fore.

Other proposals made in this report – such as 'redistributing' working time so that everyone can find paid employment but no-one works more than four days a week – are also likely to be positive. Overwork reduces the time available for many of the other things in life that matter. Achieving greater balance between time spent undertaking paid work and non-paid activities will enable individuals to pursue a range of interests and follow pathways which are intrinsically as well as financially rewarding to them.

By creating the conditions which help to promote people's well-being across its various dimensions, we can in turn expect to see measures of personal well-being, such as life satisfaction, increase.

Figure 4 illustrates how average life satisfaction in the UK might be expected to rise from now until 2050 if it was to reach Danish levels.

In this first part of this report we sketched out what life might look like after the Great Transition. We also made a case for why this is desirable from the perspective of our long-term well-being and set out the stark case for why such a radical change of direction is an urgent necessity.

Finally we hope to have whetted your appetites a little. By showing that if we change our focus from GDP growth to growth in real social and environmental value, we can all be better off in the ways that matter. In so doing, our well-being will have the space to flourish and to thrive.

Figure 4. UK life satisfaction in Great Transition scenario.

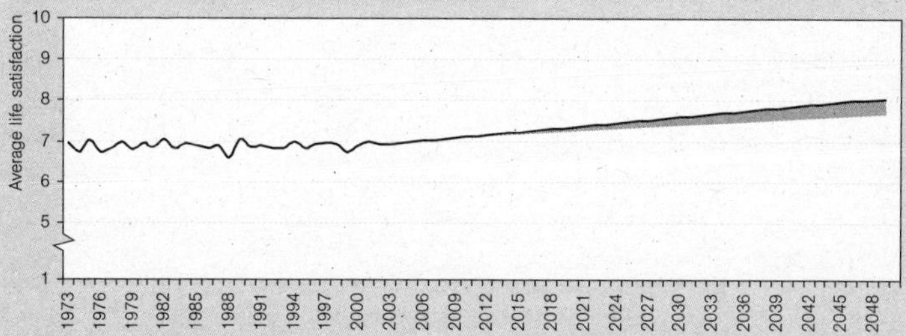

Note: Shading indicates margin of error

Part 2: The Great Transition

A map of the world that does not include
Utopia is not worth even glancing at.

Oscar Wilde (1854–1900)

The Great Revaluing

I perceive that the great part of the miseries of mankind are brought upon them by false estimates they have made of the value of things.

Benjamin Franklin (1706–1790)

The success of our economy has always depended not just on the size of our gross domestic product but on the reach of our prosperity; on the ability to extend opportunity to every willing heart, not out of charity but because it is the surest route to the common good.

US President Barack Obama, Inauguration speech, 2009

What is important to us? What do we value? These are questions that as a society we do not ask often enough. In the previous section we set out a tantalising scenario for how we might move from the partial and sometimes self-defeating goal of GDP growth to embrace a richer notion of value and progress.

Intuitively we know this makes sense. Governments as far back the Kennedy administration in the USA have recognised the short-comings of GDP, yet it still lives with us as a headline measure of progress. The reason that GDP is an attractive measure is that it counts things that are priced and therefore easily measurable. This is ultimately perverse because it is made up of lots of things that we can count – including prisons, pollution and weapons – rather than things that matter, such as caring and friendships. This is not just an intellectual curiosity; what we measure and value doesn't just describe the world, it also makes the world. It influences our behaviour and determines how we organise ourselves as a society, prioritise our activities and allocate our scarce resources.

This kind of thinking at the macro-level has cascaded into other areas of life distorting behaviour in a similar way. The critique holds whether we are talking about the public or private sphere: short-term financial costs and benefits are paramount. Public companies are under pressure to deliver ever-increasing returns to shareholders in an ever-shorter timescale and governments seek to balance budgets and boost economic growth within short fiscal and political cycles. In this situation, things that are market-traded and deliverable within the short-term take on a far greater significance than they perhaps merit. The outcome is institutions dominated by incentives that don't have the creation of social value as a central aim.

As far back as Adam Smith, economists have struggled with the paradox of value, where diamonds are socially useless, yet expensive and water – essential to our lives – is infinitely cheaper. If we rely on the unregulated market to price everything then we will find ourselves in the ultimately self-defeating situation where we only value life-giving water once it has become scarce and polluted. The effects of this can be seen on the non-market economy, or the core economy.[35] The core economy is made up of all the unpaid labour that is crucial to keeping society and communities functioning – the time that is put into caring for older people, raising children, volunteering for the community, exercising democratic rights to protest, and raising children. These tasks (disproportionately carried out by women) have also been systematically devalued as the market economy encroaches on the time that can be committed to them. Where markets have emerged in these activities they are poorly rewarded, making up the notorious '5Cs' – cleaning, cooking, care, clerical and cashiering.

This leads us to the final distortion created by the way in which we value things. The current approach is also skewed towards the interests of the already powerful. The design of our society reflects the way in which we value things. Therefore it should come as no surprise that the power imbalances we see elsewhere are evident here. In the production of goods and services, the costs to the people who are most directly affected are often invisible. Our prisons are full of people who have the least to lose in society – the poor, the mentally ill, the disabled and the drug-addicted – whilst those that perpetrate other kinds of less visible social harm are often not held to account.

For all these reasons, a process of revaluing – the Great Revaluing – is central to the Great Transition. We have placed this first deliberately, as much else depends on getting this right. In designing a new economy, we would initiate a process to comprehensively map out what we collectively value. This could take the form of a national census that explores these issues. This doesn't require starting with a blank sheet of paper. Our well-being research tells us it is rooted less in material goods and more in the quality of our relationships and the meaning we get from our work. But we have other, more difficult questions to answer.

In the same way that the market is not the arbiter of all value, neither are market-traded goods worthless. This means unlearning much of what we feel we know about value and educating ourselves on the subjectivity of it; embracing the idea that as the French author, Moliere, once said, things only have the value that we give them. This Great Revaluing needs a common understanding of value that is based on what we deem collectively, as society, to be important.

The collective is key here. Value in this sense is determined not by what we each want for ourselves – where this might be something that impoverishes another – but by what we agree is important for all of us, as members of society, to have access to – such things as a functioning ecosystem, a right to safe shelter, access to food and water and, ultimately, well-being. It also involves acknowledging that we will have competing interests that require government leadership to manage. We will have to become comfortable with trade-offs if we are indeed to redefine value in a meaningful and egalitarian way.

Measurement of well-being would also become central to our interpretation of progress. If actions are directed towards achieving better outcomes and are successful in doing so then a rise in aggregate life satisfaction would follow, as well as improvements on more detailed measures of different well-being components, such as the quality of our relationships and mental health. For this reason, comprehensive measures of subjective well-being, such as the National Accounts of Well-being (NAWB) pioneered by **nef**, would be put in place at the outset of the Transition and reported on annually, alongside adjusted measures of GDP such as the Index of Sustainable Economic Welfare (ISEW).

Pursuing social value can also influence behaviour. The core economy would be given the space that it requires to grow. Within the market economy we would reward useful jobs and penalise destructive ones. The views of people who are most closely affected by the production of goods and services would be systematically included in order for their outcomes to be fully incorporated. We would seek to reconnect stakeholders with policy decisions that affect them, rather than construct abstract models that are a triumph of mathematics over reality.

In the Great Transition, this socially defined concept of value is placed at the centre of decision-making and progress towards it is measured. In public policy, achievement of such value would be instituted as the central goal. Instead of living with the consequences of spiralling inequality and climate change whilst politicians are applauded for achieving economic growth, politicians would be held to account on the social and environmental as well as economic outcomes that they achieve. Policy decisions would be audited across the 'triple bottom line' using approaches such as Social Return on Investment (SROI).[36]

But it is not just in the public sector that this new socially defined concept of value would have an effect. Involving businesses makes this potentially transformative. A 'great repricing' that aligns prices with value would be instigated. This would require businesses to take full account of the costs of any (unintended) negative consequences. Through intelligent use of the tax system, the price paid by the final consumer would be aligned with real value.

Take the example of wood furniture manufacturing. At the moment, a manufacturer is not penalised for sourcing wood unsustainably. They pay for the wood, make their chairs and put a price on it based on whatever their costs were and the margin they want to make. If we value forests and require anyone that buys unsustainable wood to pay for the environmental costs of deforestation, then the price of the unsustainable wood and/or the furniture made from it would go up relative to sustainable wood sourced. Similarly, food grown locally is often more expensive than that shipped around the world, not least as the environmental costs of the transport involved are not reflected in the price. Repricing in line with what we value would fundamentally change incentives for consumers and so for business – there would be a significant disincentive to activities that lead to social or environmental harm and an incentive to do good as this would lead to greater price competitiveness. Successful companies would be those that build the most social and environmental value. Those that are destructive of this value would not be able to compete, forcing them to change their ways or to go out of business.

We are already seeing the beginnings of this with carbon pricing, which shows the power of prices. By aligning prices with what we value we ensure that they are harnessed to act in the interests of society and the planet, rather than in the interests of a powerful few.

Deciding on the things we value is one thing, how these should be distributed is quite another. The Great Transition described in this paper is one where outcomes should be distributed relatively equally. We can think about this in a number of ways. On the one hand, there are objective 'hard' individual outcomes, such as life expectancy, physical and mental health, or material living standards. Socially, equivalents would be the standard of housing and transport infrastructure, the availability of fulfilling work, opportunities for learning, or the quality of local public services.

The Great Redistribution

People claim that they are indifferent to the distribution of resources at any one point in time, as long as everyone has an opportunity to achieve success are effectively deluding themselves. Equality of opportunity often requires an equal starting point.

OECD, Growing Unequal, 2008

Between persons of equal income there is no social distinction except the distinction of merit... That is why idiots are always in favour of inequality of income (their only chance of eminence), and the really great in favour of equality.

George Bernard Shaw, playwright (1856–1950)

Nobel Laureate Amartya Sen looks at it from a different perspective and asks if the 'capabilities' required to enable a flourishing and fulfilling life are met across the population in a relatively equitable way?[37] The result of this would be likely to be an equitable distribution of the outcomes described above, though this is by no means guaranteed of course. Generally, however, if we have equal capabilities, we will have relatively equal outcomes. If we can do this, we will have created the conditions for the well-being of all, both individually and socially, to flourish and can turn our attention fully to developing quality rather than quantity in a post-Transition society.

To achieve this, the Great Transition would see redistribution in three areas: income, assets and time, and would seek to extend this to a fourth – carbon.[38] The first of these is *incomes*, which have become increasingly unequal in the UK since the 1980s. This has had very damaging effects, individually, socially, economically and environmentally. At the individual level, well-being suffers in unequal societies, as relative status is an important determinant. This is also associated with higher consumption, as people tend to aspire to the consumption patterns of those on higher incomes than themselves, stoking materialism and pulling total consumption up. The environmental consequences of this are all too obvious, but in social terms, income inequality is also highly corrosive for community cohesion, which is associated with higher levels of social ills, such as crime, for example.

In the first part of this report we illustrated a process of income redistribution in five stages, which would reduce income inequality in the UK to that seen in the most equal European societies, such as Denmark. A genuinely progressive tax system, phased in over 10–20 years would achieve this. The advantage of such a Transition would be that taxes could then be shifted towards 'bads', such as environmentally damaging factors, rather than 'goods' such as income and employment. This will be described in more detail later in the report.

Key to preventing the generation of income inequalities in the first place is the distribution of assets or wealth. This is the second stage of the Great Redistribution. Assets such as land, property and family wealth are even more unequally distributed in the UK than incomes.[39,40] Indeed, the possession or lack of such assets is itself a major determinant of incomes. Wealth enables a good education to be purchased, either directly through the public school system, or indirectly through the ability to buy a house in the catchment area of a high-performing state school. The result of this is that the UK is one of the least socially mobile countries in the developed world.[41] The children of the wealthy go on to be wealthy themselves. The children of the poor stay poor.

To give everyone an equal chance in life, we need to level the playing field in terms of the ownership of *assets*. There are many ways this could be done, but one straightforward approach would be to provide all citizens with a cash transfer on reaching adulthood, money which could be used to start a business, or could be invested in real assets in their local communities, or in human assets such as continuing education. If of sufficient size – we propose £25,000 per individual – such a 'Citizens' Endowment' would both level the playing field and move us towards a genuinely 'stakeholder society'.

The logical way of funding such a transfer is through a phased increase in inheritance tax, which would both generate the sums needed and itself start to reduce the unequal distribution of the relative assets available to young people in our society.[42]

A society in which all have a real stake is about much more than financial assets. As inheritance tax rises, this would see more land and property being transferred to the state, which would form the basis for redistribution on a non-individual basis. These assets could form the basis of community land trusts,[43] where land is commonly owned and managed on a stewardship basis by the local community underpinning the provision of affordable or social housing.

The Government of Margaret Thatcher envisaged a stakeholder society of sorts, where shares in privatised state enterprises would be widely held throughout society. Despite its obvious problems, there was the germ of a good idea hidden in there. What would make much more sense would be to move to a far greater level of employee ownership, where publicly listed companies would progressively transfer shares to their staff, giving them a real stake in and control over the companies where they work. With similarities to the Meidner Plan[44] in Sweden in the 1980s, such a process could lead to the creation of a series of co-operatives, operating in regulated markets, and subject to competition from new companies, perhaps also formed co-operatively as young people pool their Citizens' Endowments to start new companies.[45,46] Such a process would fundamentally change power relations within workplaces, creating a form of economic democracy.

The final aspect of the Great Redistribution is that of *time*, or more precisely, *work* and *time*. Just as income and assets are very unequally distributed in the UK, so too are work and time. For those on above-average incomes, working hours per week are far in excess of other European countries. At the other extreme, one in five households has no one at all in employment. Astonishingly, three-quarters of single mothers from low-income backgrounds have no income from employment.

Something is very wrong in a society that suffers simultaneously from chronic overwork and chronic lack-of-work, both of which are highly corrosive at personal, family and social levels. By redistributing working hours at the outset of the Great Transition we address income poverty at the bottom and time poverty at the other end of the income distribution.

Even on average we work too much. Working hours in the UK exceed those in most other developed European countries, not least because of the UK opt-out from the *European Working Time Directive*, which allows companies and workers to 'agree' that working more than the Directive's 48 hours per week is acceptable. The final part of the Great Redistribution would see total permitted working hours gradually reduced from 48 to 32, equivalent to a 4-day working week for all. In time, it might be possible to reduce this further, perhaps even to 21 hours, or a 3-day week. How would these changes impact upon our sources of value described above, as well as human well-being?

Well, as consumption – and therefore economic output – is reduced to meet the UK's carbon reduction obligations, this would help ensure that unemployment did not increase. As already pointed out, the equal distribution of working time would contribute significantly to relative poverty and inequality, and to the reduction of social ills and so the creation of the sort of society that people would actually want to live in. Finally, equally distributing increased time to spend with family (caring or just being together), friends, or in leisure pursuits creates more space for individual well-being to flourish, as well as for more community engagement and so the advancement of social cohesion and the growth of the core economy.

Just outcomes can also be good outcomes for individuals, for communities and for society.

The Great Rebalancing

As Adam Smith noted, freedom of exchange and transaction is itself part and parcel of the basic liberties that people have reason to value. To be generically against markets would be almost as odd as being generically against conversations between people.

Amartya Sen, *Development as Freedom* 1999

So far we have looked at what we value and why, and how these things should be distributed through society so as to ensure that all people have the assets needed to live a good life where their well-being can flourish.

Another way of looking at these parts of the Great Transition is that we have laid out the boundaries of the playing field, ensured that the 'players' are equally well equipped and taken care that the field itself is level. This says nothing about what happens when the whistle blows to start the game. In particular, how are the players to interact, and what are the rules of the game?

Classically, the divide in terms of economics is between central planning at one extreme and free markets, or laissez faire, at the other. To continue our analogy, does the referee assign roles and stipulate who does what and when (central planning), or does each player follow their own interests and play their own game? As a spectacle, neither would be particularly appealing: predetermined stasis or uncoordinated anarchy.

In reality, of course, you want a bit of both, which is why most economies have been mixed economies to varying degrees. Control in terms of a guiding strategy for the players, which sets overall objectives and the style of play. Freedom in that players can innovate within this framework, respond to changing events and do the unexpected, and cooperate with each other in pursuit of their shared goals.

Returning from this rather tortured analogy to our Transition, the parallels are clear. How should an economic system balance control and freedom, and what role should cooperation play in both the processes of economic and social life, and in setting the framework – or the rules of the game – in which this takes place?

The role of markets

Let's start with the question of freedom, specifically its relationship to markets. We have already argued that a key part of the Great Transition will be for the market prices of goods and services to reflect the social and environmental costs (and benefits) associated with them – they should incorporate both negative and positive externalities. The fact that this has not been the case makes it clear that our so-called free markets have been far from free. Rather, many people's freedoms are severely curtailed. Whether this is the freedom to breathe unpolluted air or to drink clean water, or the freedom that comes from getting a living wage for your work or a fair price for your produce, the outcome is the same: our freedom to choose is restricted by the fact that 'bads' such as pollution and poverty-level wages are not reflected in market prices and so are far more prevalent than they should be.

In the same way, the freedom to buy 'goods' such as locally produced organic food or ethically sourced clothes is restricted by the fact that these are generally more expensive than alternatives without these positive attributes. Our economic system encourages the overproduction of 'bads' and the underproduction of 'goods'. At society level, this restricts the freedom of us all. By incorporating these factors into market prices we reverse the incentives: buying 'good' things becomes cheaper and easier than buying 'bad' things, and if things are *really* 'bad' we should not be able to buy them at all.

But this raises the question of why we should not just ban markets altogether. Historically, many people have drawn just this conclusion. Taking inspiration from Marx, and particularly the influence of Lenin and the founding of the Bolshevik Party, central planning in the Soviet Union was designed to ensure the fair outcomes that markets, it was thought, could not.

At least until the 1960s many thought that the centrally planned model was also inherently more efficient than its capitalist counterpart.[47] By designing out wasteful innovation and coordinating economic inputs and outputs intelligently and strategically, needs could be met more easily and living standards increased more rapidly. As it turned out, of course, this was far from being the case. Innovation was not a wasteful irrelevance but a central strength of capitalism. Why is this?

Friedrich von Hayek was an early critic of the central planning model, a fact that endeared him greatly to later free market 'revolutionaries', not least Margaret Thatcher. Hayek was actively involved in the 'socialist calculation debate' in the 1920s and 1930s, which sought to determine how best to allocate resources in a society.

Socialists argued that a central planner with sufficient information about the needs and wants in a society and the resources available to achieve these, was best placed to do this optimally. Hayek, in something of a minority, argued that this was effectively impossible, and attempting to do it would lead inevitably to totalitarianism.

Hayek focused on the constraints imposed by the 'knowledge problem'. Simply, this stated that the amount of information needed to coordinate activities across an entire economy was practically impossible to gather, even if it could, in principle, be computed. Economists, quite naturally, have tended to look at this in terms of allocating and coordinating the 'factors of production' (land, labour and capital), but Theodore Burczak emphasises a quite different aspect of Hayek's thought.[48]

Far from being one of the fathers of neoclassical economics with its rational, computer-like 'utility maximisers', Hayek stressed the importance of subjectivity and the fact that our knowledge is inevitably partial and 'bounded'. We create and live in our own worlds. While we might be able to get a sense of what others need, it is not possible to know what others *want* at any point in time in any meaningful sense.

Central planners cannot get this information from inside people's heads. The result of a centrally orchestrated allocation process will thus be the interpretation of a small group of people of what an entire population *might* want. At best this might cover basic needs, but even this is not guaranteed, and it cannot but be strongly influenced by the whims of the planner.

If it is impossible to do this at any one point in time, it is even more impossible to do so over time. As societies change, what they value changes. As technology changes, what is possible changes. For Hayek, allowing resources to be allocated through the price mechanism operating in free markets was the only means of resolving these problems.

In principle, at least, the price of a thing should reflect its scarcity, the demand for it (i.e., how much people value it and can afford it) and its attractiveness relative to other goods. Through the interaction of countless separate transactions, markets should allow the 'discovery' of a price that balances all of these components. The way that prices change over time then reflects changes in the balance of these aggregated individual decisions, reflected in price levels.

What does this mean in practice? Well, 40 years ago a computer was the size of a car and more expensive than a house. Today, personal computers are hundreds of times more powerful, are seen as essential by most people in developed countries and cost less than the average industrial weekly wage in these economies. What was once seen as a technical tool for specialists turned out to be a thing of very high value for a large proportion of the population. It is very difficult to imagine how a centrally planned economy could have arrived at this outcome.

Similarly, today, everyone in developed countries – and increasingly in developing countries – has a mobile phone. Mobile phone companies emerged to challenge the established fixed-line providers, based on the belief that there would be a market for their products. Would a central planner have tried this? Would even the fixed-line providers have done so? Innovation and the ability of entrepreneurs to test hypotheses about what people might want have clear benefits for society. The alternative is to just produce the things that have been produced in the past, or for some central agency to try and predict what people might want and then to impose it upon them. Neither seems particularly attractive.

Problems with markets

There are clear problems with markets, which have long been recognised. Managed or 'social' markets have sought to regulate, temper and steer the market so as to reduce negative social impacts and prevent market economies becoming 'market societies'. In the UK, the 'third way' attempted to combine the logic of free markets with the public service ethos and the energy of civil society. None of these often laudable attempts have really succeeded, and we would argue that they could never do so until we 'get the prices right'.

This is simply unsustainable, and no amount of tinkering with regulation or appeals to business leaders to take a triple bottom line[49] approach has been sufficient to offset the logic of the market expressed through the single bottom line of profitability. In the Great Transition this would be decisively resolved. By using the tax system so that 'goods' become cheap and so plentiful, and 'bads' become very expensive and so very rare indeed, the strength of the market is harnessed towards positive goals, rather than constantly undermining them.

This is not the only problem, however. The argument set out above might suggest that the things people want is a sort of natural phenomenon just waiting to be uncovered by the market. If this was the case, why would companies bother spending billions on advertising products? Clearly, 'wants' can be and are manufactured and manipulated. Consumption has been stoked to unsustainable levels, becoming an end in itself for many. The Great Transition would inevitably see overall consumption reduced, and strict restrictions placed on advertising, particularly to children. At the very least, advertisers would be required to prominently feature the social and environmental costs and benefits of their products, but the fact that these would also be incorporated into market prices would limit the extent to which the advertising of 'bads' could succeed anyway.

In the Great Transition prices would still be used to allocate resources, but by making them reflect real value, we would create a race to the top where resources are used to maximise long-term social and environmental value, not short-term profits made at the expense of our society and our natural environment. Prices would operate within the parameters set to meet explicit social and environmental goals.

The role of the state

The state is 'us' not 'them' – or at least it should be. It is not a malign force or an alien entity, but the product of collective decision-making and an expression of shared interests. It is the means whereby we meet needs that we can't meet individually and the source of defence for relatively powerless individuals and groups who would otherwise be overridden by more powerful interests. It is, or could be, what we make it.

In terms of the state's role in the economy, however, it could be argued that getting the prices right would greatly reduce if not eliminate it. Much of the justification for state intervention in the economy comes from the fact that the market does not take account of externalities.[50] Because of this, the state has long been seen as having a duty to prevent public 'bads' – like pollution – and to produce public 'goods.' But if market prices reflect social and environmental costs and benefits accurately, 'bads' become expensive and are produced less while 'goods' become cheap and are produced more. Would we still need the state at all then?

In the Great Transition the unequivocal answer is 'yes'. There are four reasons for this.

First we should distinguish between public 'goods' as described above, and 'public goods' as understood in the strict economic sense. The way the term has been used thus far relates to things that create positive social and/or environmental value in a broad way; essentially anything that most people would agree was positive for society or good for the environment. In economics, the concept of public goods means something much more specific than this.

Paul Samuelson is generally credited with developing the public good concept in economics.[51] Samuelson identified two characteristics of any such good. The first is that it is 'non-rival', which means that my consumption has no bearing on the ability of others to consume. For example, no matter how much air I breathe this does not reduce the amount available to others. Second, it is 'non-excludable' in that it is practically impossible to prevent people from consuming the good. Again, air is a good example of this. Samuelson pointed out that, because of these characteristics, individuals or private companies have little or no incentive to produce public goods, as they are not able to capture the benefits of doing so.

An example often cited is street lighting. If we live in a street without lights, no individual has sufficient incentive to pay for their installation, as everyone else in the street would reap the benefits (and could not be prevented from doing so) despite the fact that they had not contributed. From an economics perspective, they are 'free riders'. This concept can be applied to much more than street lighting.

We can think of the rule of law as a public good, for example. Similarly, maintaining broad economic and financial stability is a public good that is essential to underpin local, regional and national economies. Everyone benefits from this, but no single individual or group of individuals could or would maintain the rule of law or macroeconomic stability across a society or an economy. In a related way, maintaining reasonable levels of individual and regional equality can be seen as a public good – the negative effects of high levels of inequality extend beyond individuals affected to society as a whole.

Because of these factors public goods in a market economy will be undersupplied. A classic argument for the role of the state is thus to fill this gap and produce the public goods that we would otherwise lack, and which in many ways are the hallmark of a real society that is more than the sum of its cities, towns, villages and neighbourhoods.

So, if we have addressed the issues of externalities and public goods in the Great Transition, should we then leave it at that? The answer must be 'no'. There are further issues to consider, which turn on the relationship between the state and individuals, and more broadly between the state and society as a whole – at national, regional and local levels.

The state and individuals

Earlier we examined the reality of freedom and choice in the context of economic externalities and argued that for most people, real freedoms are restricted and many choices are illusory. But even if we can address these issues, there remain serious tensions with regard to choice. There are two parts to this: the individual and the social.

From the perspective of the individual, people don't always make choices that are in their long-term interests. The choices we make are often influenced by circumstances over which we have little or no control – and the less knowledge and power we have the more heavily edited our choices are likely to be. We also have a tendency to overweight short-term gratification at the expense of long-term well-being. The decisions we make, however constrained, set us upon paths, and paths taken at a relatively early stage in life can have major influence upon later well-being. It can also be fiendishly hard to change direction once a path has been established, though this is not impossible. Coaching relationships, if conducted well, can provide a space where options can be explicitly explored and long-term choices made on a different basis.[52] For most people in most circumstances, however, choices are not made in contexts like this.

Of course, people need to be genuinely informed about the determinants of well-being, but that's not enough. Decisions that are 'right', according to the best available evidence, can still be hard to make.[53] People discount the future heavily, even when they have full information about costs and benefits. We are very bad at imagining how we will feel in the future and estimating how likely future events actually are. Carefully designed regulations and institutions have a key role to play here. While the state cannot make people happy, it can create frameworks and help to shape the circumstances that make it easier for people to take decisions that are positive for their long-term well-being.[54] To make the 'right choice' the 'easy choice'.

That said, the right choice for one person may not be so for another. While there is evidence on what sort of choices are positive for our long-term well-being, we cannot be too prescriptive about this – we can perhaps 'nudge', but not 'shove' people in any particular direction, and need to be very careful not to close down the options that people face.

But let's be clear: this is about much more than encouraging individuals to change the way they think and behave. If we are all to have a fair chance of choosing things that will bring us long-term benefits, then it is only through the collective mechanisms of the state that steps can be taken to ensure that everyone has access to the knowledge they will need to make the right decisions, and has sufficient power (in economic, social and political terms) to apply that knowledge, and is able to participate fully in the processes that make and implement decisions. Making the right choice the easy choice for everyone requires systemic change to redistribute power and resources across the population – the very process we have described in the Great Redistribution.

The state and society

For the Great Transition, we shall need the state not only to enable individuals to make choices that serve their own long-term interests, but also to promote and protect the long-term interests that we share. For example, we share an interest in having access to healthcare when we are ill and to education for our children, in feeling safe on our streets at night, in safeguarding the natural environment. We need these things for tomorrow as well as today. It would be well nigh impossible to meet these needs on an individual basis. So we must be able to meet them collectively through the state and at the same time to safeguard the capacity of the state to do so for the future. This will require a fundamental shift in how it sets priorities and deploys resources.

For a start it must give priority to preventing harm, rather than dealing with the consequences. Today's state is buckling under the strain of dealing with problems once they have arisen. Yet failure to prevent avoidable needs arising is unsustainable, unethical and unjust. Unsustainable because in a low-growth or no-growth economy, there will be a lot less money to pay for public services: funds for meeting needs that cannot be avoided should not be wasted on meeting those that could have been prevented. Unethical because avoidable risks – including obesity, mental illness, homelessness, incarceration and educational underachievement – undermine people's well-being. Unjust because the burden of risk falls most heavily on the poor.

A state that gives priority to prevention will have a different framework for decision-making – one that recognises the value of investing in upstream measures, where benefits accrue across sectors and over the longer term. It will seek to prevent 'ill-being' by addressing the underlying causes of unequal opportunity and tackling avoidable risks to physical and mental health. It will invest to prevent the waste of human potential by fighting inequality and entrenched patterns of unemployment. It will act to prevent wealth escaping from poor neighbourhoods by keeping resources circulating locally, through support for local businesses and trading within communities. And of course it must prevent damage to the environment – most urgently, climate change. There's a double prize for getting it right – more well-being for all, while public services are safeguarded for the future.[55]

The state and core economy

Today's state is based on the premise that the economy will continue to grow to yield more and more taxes to pay for more and more services. By and large, these services are designed to do things to and for people, and to fix things once they have gone wrong. The implicit assumption is that needs will multiply and services will expand. But that assumption no longer holds. We cannot rely on continuing economic growth. Where, then, will the resources come from to prevent harm and to deal with the consequences when harm cannot be prevented?

For the Great Transition, the state must combine a refocused tax system (see the Great Economic Irrigation) with ways of releasing those resources that exist in abundance in what we have called the 'core economy' – the wealth of human and social assets that are largely neglected by today's public services. These are embedded in the everyday lives of every individual – time, wisdom, experience, energy, knowledge and skills – and in the relationships between them – love, empathy, watchfulness, care, reciprocity, teaching and learning. They make the world go round. They are mainly unpriced and unvalued. In large part, they have been deployed by women – not because they belong naturally to women, but as a result of long-standing gendered divisions between paid and unpaid labour. What the state must do now is to nurture these resources by valuing them and supporting their development. By redesigning employment policies, childcare, and family support, as proposed in different parts of this report for example, it can help them to flourish and grow.

The core economy is plentiful, dynamic and sustainable. Growing it shifts the focus of state activity from doing things to people, to acting with them; it shifts the emphasis from taxing and spending to human agency and social exchange.

Co-production is central to this approach.[56] It goes well beyond the ideas of citizen engagement or service user involvement to foster the principle of equal partnership. It offers to transform the dynamic between the public and public service workers, putting an end to 'them and us'. Participants are no longer either providers or users. Instead, people pool different kinds of knowledge and skills, based on lived experience and professional learning, to co-produce well-being for all.

Co-production puts people in touch with each other: it strengthens social networks and helps to prevent ill health and improve services. Becoming equal partners or co-producers instead of users can transform people's relationship with their own well-being, as well as their experience of public services, because it enables them to feel valued and to have more control over what happens. At the same time, by working in equal partnership with those they are supposed to serve, public services can dramatically increase their resource base, radically transform the way they operate, and achieve better results for the people they serve.[57]

The state, neighbourhoods and society

While it is basic for human beings to trade and exchange, it is just as fundamental to cooperate. We are social beings who are at our best as active participants of thriving and vibrant groups and networks. Not everything can and should be traded. Carving out and maintaining a non-market, social sphere is thus an essential part of the Great Transition. This involves valuing and nurturing the core economy. But there is more to it than that. Places matter: the character of neighbourhoods, the employment they offer, the quality of local services and amenities, how people interact with each other in streets, parks, shops, pubs and other meeting places. And above all what matters is how much power people feel they have to shape the places where they live and to alter or conserve them to suit themselves and each other.

There is a key role for the state in distributing resources between localities and regions so that every neighbourhood has a fair share of employment, services and amenities. It must devolve power so that people can exercise as much control as possible over decisions that affect them locally. It must also promote equity between localities and regions – not uniformity, but common standards in meeting shared expectations and needs. The move must be towards greater devolution and local empowerment, within a framework of universal rights protected by law. This is not an easy balance to strike. Two strategies should help.

The first is to build an increasingly strong culture of shared interest. Political leaders can tell stories to the electorate that over time shape people's sense of themselves and each other. If governments could spend the last 30 years promoting competitive individualism and a wild enthusiasm for market solutions to social problems, why not spend the next few decades promoting the benefits of cooperation and interdependence? Children in Scandinavian countries tend to be seen as a 'public good' for which everyone bears some responsibility.[58] In the Great Transition we would attach a similar meaning to childcare, education and health.

The second – and most important – strategy is to open up a vibrant, inclusive, informed, creative, democratic debate at national as well as regional and local levels. Its purpose would be to negotiate the tricky questions that arise when local empowerment comes

Case Study 1. Blaengarw – co-producing a vibrant community in the Welsh Valleys

The community of Blaengarw is a small and isolated former mining village near Bridgend in South Wales. The Blaengarw Workmen's Hall was built in 1894 and was at the heart of the mining community in the Valley and the centre of educational and social activity. 2,500 miners and their families working in the Garw Valley used the Hall on a regular basis. Since the closure of the last six mines in 1984, the Valley has seen a decline in the number of residents, a large increase in unemployment and a slow disappearance of the strong traditions of mutualism and collective community identity centred on the hall.

In recent times, however, the Blaengarw Workmen's Hall has been revitalised. The 'Time Network', which officially began in April 2005, enables members of the community to earn credits by giving their time to help with running the community groups in the area (e.g., youth clubs, the rugby club, the hall, festival groups, churches, residential homes, the community café). Members of the Time Network use their credits to access social events and activities at the hall or in the other venues in the community (social club or rugby club).

The Blaengarw Workman's Hall Time Centre now issues actual paper currency in denominations of one or two time credits. To attend the events in Blaengarw individuals can either pay in cash or with time credits. The time credits are always strictly hour-to-hour; whether it is bingo or opera, a three-hour event is always three time credits. The cash amount is always the market value of the event, so attending the bingo would be cheaper in cash than attending the opera.

The Creation café is also an important part of the Time Network. Members earn credits by giving their time to support the paid staff running the café. Credits can be used at the café; for example, an hour on the internet can only be accessed with a time credit.

In this way, Blaengarw has effectively established its own complementary community currency which serves to increase local spend within the community, benefiting the local economy as well as building social capital.

Over the last few years, membership of the Time Network has grown to 500 people who all give their time to the community and access the events and activities. New groups have developed as individual members with similar interests have connected through the time network. The community has also come together for larger projects; for example, in summer 2007 Blaengarw held its first festival for 21 years.

up against universal needs and expectations, at a time of shrinking public resources. We expect the state to play a significant part in the Great Transition, but for that to work we shall need a radical overhaul of the democratic structures and mechanisms that underpin the state. Taking the co-production approach, where everyone is valued and everyone participates, we will ensure that the state really is an expression of our shared interests and goals.

The Great Localisation and Engagement

I sympathize with those who would minimize, rather than those who would maximize economic entanglements among nations. Ideas, knowledge, science, hospitality, travel — these are things that of their nature should be international. But let goods be homespun wherever it is reasonable and conveniently possible, and above all, let finance be primarily national.

John Maynard Keynes, *On National Self-Sufficiency* 1933

The financial crisis revealed in very stark terms the limits of globalisation. Huge, multinational investment banks, with turnovers many times larger than most of the world's nation states, imploded, threatening the stability of the entire global economic system and eventually having to be propped-up by governments at massive cost to taxpayers. There seemed no alternative to but to save them; they had grown too big to fail.

But the globalisation of finance is just one example of the problem of scale in modern economies and modern economic theory. The emergence of clearly visible environmental limits, the instability, inequity, alienation and frequent resource inefficiency of global markets and the more subtle erosion of the community and social networks that large and unaccountable corporate bodies create, have all conspired to bring the problem of appropriate economic size in to sharp focus.

The Great Transition will require and lead to a Great Localisation. This will have the principle of subsidiarity at its heart. In political terms, subsidiarity is widely accepted as meaning that everything is best decided and implemented closest to the people who will be affected. It is a central principle of the European Union, if not always the practice. When discussing the economy, we can argue that economic activity should be conducted at a scale and over an area that maximises social, environmental and economic benefits and minimizes costs as described in the Great Revaluing section. We should find the right scale for our needs, not assume that bigger is always better.

When we talk about localisation and scale then, we are not simply saying that everything should be smaller. *Appropriate* size is what we are looking for in the Great Transition. Economies of scale are vital for efficient production in some areas. It makes no sense for every town in the UK to produce its own computers. But we have gone too far the other way, and assumed that nothing can be produced efficiently on a local scale. This obviously turns on what we mean by efficiency of course, but the key point is that just as there are economic benefits to producing some things on a large scale, there are major environmental, economic and social benefits from doing many others things on a more human, local scale. There are also very big costs associated with relying only on the large scale.

The financial crisis has demonstrated that greater size creates greater vulnerability to economic shocks. With peak oil imminent and climate change already upon us, the Great Transition will involve a major shift away from the energy-intensive production processes involved in getting food on to our plates, for example. During the truck drivers' strike in 2000, then Prime Minister Tony Blair was informed by the Chief Executive of Sainsburys that the country was around nine days away from food rationing and civil unrest.[59] The average UK citizen now travels 893 miles each year to buy food.[60]

Resilience, efficiency and diversity

Resilience can be understood as the 'capacity of a system to absorb disturbance and reorganise while undergoing change'.[61] New Economics recognises that there needs to be a balance struck between production efficiencies, the scale of a system and the ability of an economic system to adapt to shocks.[62] Technically it may appear to be efficient to have a large provider of goods or services who is able to achieve a scale of production which lowers per unit costs. This is a familiar retail model in the UK, as we have seen with the rise of the supermarkets and their highly centralised distribution system.

However, a local economy entirely dependent on this retail model is considerably less able to weather economic shocks when compared to one with a diverse range of smaller retail outlets, sourcing products across a range of different length supply chains. At the local level, an economy with a mixture of ownership models – e.g., co-operatively or family owned, social enterprises, small privately owned businesses – demonstrates a diversity not only of ownership, but a diversity of motivations for doing business in their local area which in turn strengthens the resilience of the local economy. Larger retail outlets, when faced with an economic shock like today's recession, will simply close those stores that are less profitable, with little concern for the consequences, much as Marks and Spencer and Waterstones did recently in closing 47 stores across the UK at the cost of over 1400 jobs.[63] We see the smaller businesses weathering the storm.

Diversity in terms of size and type of commercial operation is thus important, not just because it makes our towns more attractive and is likely to strengthen community relationships; it is also vital in terms of economic resilience and sustainability. We need to start thinking about efficiency in a far more holistic way, one which takes full account of the range of things people value in life.

Our current approach to economics ignores this balancing act and offers a poor guide to appropriate scale, leading naturally towards 'gigantism' as the solution to every problem.

Trade, comparative advantage and market power

Trade is good example of how our current system creates inappropriate internationalised economic activity that could be done much more efficiently and at much less environmental and social cost at a smaller level. All around us still, are ships, lorries and planes passing in the night, wastefully carrying often identical goods from city to city across the globe and back again to meet consumer demand (Box 3); **nef** calls it 'boomerang trade'.

One strong argument for free trade is the law of comparative advantage, developed by the economist David Ricardo.[64] Put simply, Ricardo argued that it makes more sense for countries to specialise in the production of a certain good or goods and trade these rather than try to make everything, even if they can make all goods more cheaply than a trading partner. This is because opportunity costs are reduced by specialisation – two countries specialising and trading can produce more of the same goods in total than two countries that do not.

But Ricardo's theory rested on the assumption that capital would remain immobile between nations, as it was when he was writing. However, if capital can move freely between nations, it would have no reason to be satisfied with a mere comparative advantage at home, but would seek what Herman Daly has described as 'absolute advantage'[66] – the absolutely lowest cost of production anywhere in the world. With social and environmental externalities excluded from the pricing system, this leads us to a race to the bottom, with multinational companies dumping these 'bads' on

Box 2. Ecologically wasteful trade

In 2008, the UK exported 4,400 tonnes of ice cream to Italy, only to re-import 4,200 tonnes. We imported 22,000 tonnes of potatoes from Egypt whilst exporting 27,000 tonnes back again. We sent 5,000 tonnes of toilet paper to Germany, and over 4,000 tonnes returned. Similarly 10 tonnes of 'gums and jelly' sweets went back and forth to Thailand. And, at the last count, 117 tonnes of 'Sweet biscuits, waffles and wafers, gingerbread and the like' (to use the category used by trade statisticians), came into the UK, rumbling passed 106 tonnes headed in the opposite direction.[65]

those countries with the lowest labour costs and weakest social and environmental regulations.

While such a race to the bottom may well, in the end, reduce the price of consumer goods to its lowest possible level, there is nothing 'efficient' about this in any meaningful definition of the word. As well as the environmental consequences, the social impacts are profound as individuals, communities and even countries become increasingly atomised and specialised cogs grinding in the wheels of volatile global supply chains.

Through the Great Revaluing, the incorporation of these externalities into prices is central to the Great Transition. As transport reflects environmental costs it will become more expensive, forcing us to think long and hard about what we should make locally, regionally and nationally, and what we should trade internationally and why. As with a shift to more local food production, it is crucial that a phased approach be taken so as to allow current trading partners, particularly in developing countries, to adjust. In many respects broad economic trends are moving in this direction in any event. The heyday of development focused on exports to developed countries has passed. Hastened by the impact of the current crisis on global trade, many developed countries are looking to boost domestic demand and produce for their home markets, as well as focusing trade more on their neighbouring countries. We are arguing for something that would be quite complementary to this, but only over a phased period and only if managed as a genuine partnership with our current trading partners.

This transition would also require a much greater flexibility of approach in the global rules governing trade. Sometimes a few practical examples of how things can be done differently outweigh any amount of theoretical speculation. To give just one example of what this might mean in practice, Stan Thekaekara a **nef** Fellow and long-term advocate of progressive approaches to trade, offers a disarmingly simple solution. He suggests basic ground rules that might underpin a logical way for his part of Southern India to trade with the UK. 'We must have a robust domestic market first. The export market is for us the icing on the cake. If a product is produced in the UK, and not by a multinational, we will not compete with it. So tea I'd sell, but not Himachel apples.'

Criteria for determining appropriate scale

We need a better balance between economic self-sufficiency in some areas and interdependence in others, and a set of reasonable criteria for judging where this balance should lie. What might these include in the Great Transition?

1. **Actual need.** We do not all *need* to own two cars or ten pairs of shoes and the planet does not have sufficient capacity to produce them. Nor is it apparent that we need many of the complex financial innovations that characterise modern commercial banking. The Great Revaluation and repricing should see Veblen's 'conspicuous consumption'[67] becoming a thing of the past. Many sectors will shrink in size; for example, advertising, clothing and footwear as well as finance.

2. **The market size needed.** This would take in to account economies of scale and quality and the impact of these on price and its importance to people's lives. Here it is important to balance 'efficiency' in terms of economic production, with 'efficiency' in its broader social and environmental sense. It does not make sense for the manufacture of railway carriages, for example, or the regulation of the shipping industry and indeed aspects of finance to be conducted at the level of the town or the nation. But it is likely that the production of a very large range of everyday products and services, including food and furniture, and construction, repair and maintenance services would have greater social and ecological value if they were done at a local or regional level. Table 1 suggests minimum size units for adequate economies of scale.[68]

3. **The impact on local economic and social life.** In deprived areas of the economy, a greater degree of relocalisation of trade and production may well be appropriate, to create work and enhance well-being. The public sector should be at the forefront of this effort, as **nef** has argued over the last decade, shifting its procurement processes to favour local suppliers wherever feasible and creating 'local multiplier' effects,[69] but also tilting the national green industrial policy of the Great Transition towards these areas. Our adjusted social value pricing

Table 1. Minimum size units for adequate economies of scale.

Unit	District	Region	Nation	Continent	Globe
Size (miles)	20	100	500	2000	10,000
Population	100,000	2 million	50 million	1 billion	5 billion
Production	Food crops	Building materials	Clothes, textiles	Vehicles	Micro-chips
	Cash crops	Processed food	Small machines & components	Electronic systems	Pharma-ceuticals
	Housing	Furniture	Electronic devices	Small aircraft	Large aircraft
		Hardware	Steel	Ships	
	Energy (micro-renewables)	Renewable energy (wind, hydro, solar)	Oil, gas, coal		
	Energy-efficiency, housing retrofitting		Civil engineering		
			Books, films, bicycles		
Distribution	Fresh food	'Groceries'	Bulk commodities, e.g., grain	Oil, gas	
	Daily supplies	Clothes	Industrial machinery		
		Books			
		Cars			
		Household appliances			
		Seeds			
Services	Schooling	Universities	Insurance	Aviation	
	GP medical	Hospitals	Railways	Shipping	
	House repair	Public health	News Media		
	Restaurants	Safety	Telecom		
	Hotels	'High street' and local banking	Wholesale banking		
	Waste recycling	Buses	Electricity		
		Theatre/cinema			
		Water			

system will see European competition law naturally favour this relocalisation – with lowest price also signalling greatest positive value. A low-carbon, high well-being approach to economic development would become the norm for local regional public bodies rather than the current narrow focus on economic growth that still drives decision-making in Regional Development Agencies.

4 **The bioregional characteristics of a producing and trading region.** To develop an economy in optimal equilibrium with the biosphere it needs, as far as possible, to be aligned with the climatic, ecosystem and natural resource characteristics of its area. For example, it makes little sense to specialise in growing and trading non-native crops that can only survive with the support of unsustainably high levels of chemical and energy inputs.

The democratisation of money

The regionalisation and localisation of production and consumption could be assisted by the emergence of electronic regional and local currencies, complementing those operating at national and international levels. These currencies could provide effective economic feedback loops and further assist regional economic development in a dynamic fashion. So if a region suffers from a downturn, its regional currency will naturally depreciate against the national currency and other regional currencies, increasing within country 'export' demand and encouraging import substitution and more local production of services.[70,71]

The transaction costs associated with multiple currencies would be low as virtually all transactions could be completed electronically with smart cards, or mobile phones and other internet-enabled devices. In addition, electronic commercial barter would become more prevalent, with companies with excess capacity or inventories trading with other firms in mutual exchange networks. Already there has been rapid growth of these type of non-interest bearing credit schemes in the United States and elsewhere with the growth of the internet and since the financial crisis.[72]

Local political re-engagement

The one factor felt to cause disengagement that runs through all the strands of our investigation is the very widespread sense that citizens feel their views and interests are not taken sufficiently into account by the processes of political decision-making.

Power to the People, Report of the Power Inquiry

When two American economists, Stephan Goetz and Anil Rupasingha, carried out a detailed study of the links between Walmart Stores Inc. (Walmart) and dwindling social capital – the community cohesion and mutual support that makes neighbourhoods work – they found that if there was a Walmart nearby all the measures of social capital went down over the decade studied.[73]

Communities that gained a Walmart during the decade had fewer local charities and local associations such as churches, campaign groups and business groups per capita than those that did not. Walmart's presence was also shown to depress civic participation. Strikingly, communities that had or gained a Walmart store in the 1990s also had lower voter turnout.

Walmart and its 'big box' retail model (widely copied by UK retailers like Tesco and Asda, a Walmart subsidiary), it seems harmed not only local retailers, but also a wide variety of other businesses and professionals that served local retailers, such as banks and accountants. In the process, Goetz and Rupasingha concluded that 'The social capital they embody is destroyed, and their entrepreneurial skills and other forms of location-specific human capital are forever lost to the community.'[74]

An essential part of the Great Localisation will be for people from all parts of society to become more involved and engaged in local economic and political life. As with economic production and trade, the subsidiarity principle would be applied in the public sphere with local decision-making guiding priorities and budget decisions.

Following the Great Revaluing and the Great Redistribution, income and wealth in the UK will be much more evenly distributed, allowing the shift towards more devolved tax-and-spend policies and participatory budgeting. For people to be able to participate properly in this process, increasing levels of economic literacy is an essential prerequisite. This needs to cover how local economies work, but also how they relate to the wider regional, national and international economies. For people to make informed choices they need to be genuinely informed.

Case Study 2. Plugging the Leaks

Plugging the Leaks allows communities to take a different approach to local economic development; one that can have a more sustainable local impact than regeneration plans set at a national level. It regenerates from within and capitalises on the resources that a community already possesses. Above all it is a process that truly allows local people to set their own priorities.

Sneinton's open-air market has been trading on the east side of Nottingham for over 100 years. When the market faced closure in recent years, worried traders linked up with the *Plugging the Leaks* programme (named locally as 'Local Alchemy Sneinton') to relaunch the market managed by a locally owned social enterprise in 2006. Local traders were able to recognise the tangible regeneration benefits that street markets deliver and, through *Plugging the Leaks*, take action to retain them. The stalls are increasing and proving an ideal 'low cost and low risk' way of stimulating local business while boosting training and employment opportunities in the surrounding area.

As well as understanding money flows – ensuring that money is re-spent as locally as possible – *Plugging the Leaks'* participants can look at how other resources such as energy, water and consumer products flow in and out of a community to find enterprising ways to reduce their impact on the environment. Market traders are taking practical steps to make the market as sustainable as possible by sorting cardboard for recycling, while animals at the local farm and at the donkey sanctuary feed on fruit and vegetables left over at the end of the market day.

More info at www.pluggingtheleaks.org

Case Study 3. Participatory budgeting in Port Alegre, Brazil

Evidence from Port Alegre in Brazil has shown that deliberative processes can engage disadvantaged people who are currently far less likely to participate in formal politics. Participants in participatory budgeting in Porto Alegre had below-average income and education. The process stimulated the formation of the associations vital to civil society and the core economy.[75]

	Neighbourhood associations	Co-operatives	Regional popular councils
1986	240	0	0
1998	540	51	11

Participatory budgeting also led to many practical improvements for the poorer residents of the area, including, since 1988, a rise in:

- residences with running water from 75 per cent to 98 per cent
- sewerage coverage from 46 per cent to 98 per cent
- public municipal schools from 29 to 86

If we are to produce more things locally, however, we need to relearn much that has been lost. This is the subject of the next chapter.

Local and regional authorities will be more powerful, able to raise finance to fund transport, energy, and housing projects working with regional banks, as is already commonplace in many European countries such as Germany and Switzerland. At a more local level, participatory budgeting will gradually become the norm for the delivery of public services such as education, waste collection, street lighting and cleaning.

As more power is devolved from the centre to local and regional government, it will become easier to create new institutions which allow citizens to get directly involved in taking decisions and setting priorities. New deliberative forums can offer citizens the opportunity to become active participants in – co-producers of – government, rather than passive recipients.

According to the social theorist Jurgen Habermas, in a 'deliberative democracy', a heterogeneous collection of citizens with diverse opinions come together in the 'public realm to engage in structured debate, free of coercion, (hopefully) to reach considered judgments about an issue of common good'. Deliberative forums would provide opportunities for interested citizens to discuss local issues, from transport to housing, to healthcare and help steer local government.

The Great Reskilling

The labourer with a sense of craft becomes engaged in the work in and for itself; the satisfactions of working are their own reward... work is connected to the freedom to experiment; finally, family, community and politics are measured by the standards of inner satisfaction, coherence and experiment in craft labour.

C Wright Mills, American sociologist (1916–1962)[76]

Once, as the writer Erich Fromm put it in *To have or to be*, 'Everything one owned was [once] cherished, taken care of, and used to the very limits of its utility. Buying was 'keep-it' buying.'[77] To extend the life of an object, the owner would know how to polish, adjust, oil and repair it. With the advent of disposability, built-in obsolescence, the introduction of constant upgrades (similar to obsolescence) combined with mass advertising, the deregulated market found a way to shorten the lifespan of goods, earn more money, and come between the individual and their need and ability to maintain and repair goods.

Then, as Fromm wrote, 'Today, consumption is emphasised, not preservation, and buying has become 'throw-away' buying. Whether the object one buys is a car, a dress, a gadget, after using it for some time one gets tired of it and is eager to dispose of the 'old' and buy the latest model.'[78] The consequences: a huge amount of consumer waste and a widespread loss of everyday skills.

From making clothes to food growing, preparation and preserving, from building skills and the use and repair of everyday household goods, to even the making of music and art – simple skills that were taken for granted by countless generations have withered. A wide range of activities, goods and services was incorporated into the market as it expanded, like a cuckoo in the nest, displacing other ways of organising life. They were then commodified and sold back to the people from whom they were taken. The subtle knife of the market severed individuals, families and communities from their abilities to do things for themselves.

In the Great Localisation, we showed how trade would be radically regionalised and localised as a natural outcome of the Great Revaluing. This relocatisation will also require communities and regions to relearn and develop, combining local production and trade in the market, with local non-market exchange. In the former case, community vocational colleges providing training in the skills needed locally would be reinvigorated and linked to local businesses as apprenticeships once again become the norm.

But non-market exchange is just as important. Almost by definition, the market's expansion forecloses on the possibility of exchange based on reciprocity, mutual aid and gift giving – all of which, when left to flourish, strengthen the fabric of community and so boost the core economy.

In the Great Rebalancing we described the importance of setting clear boundaries on the market sphere so that space for core economy activities is maintained and enabled to grow. Maintaining this balance is a central component of the Great Transition.

The Great Reskilling will help end binge-consumerism in the face of environmental threats, increase the resilience of local economies and societies to external shocks, and lay the foundations of enhanced human well-being. Look around the UK today and you will see that it is already beginning. In the process communities are pulling together – as with the Transition Towns movement, for example – economic self-education is spreading and the narrow world-view of the so-called 'free market' is being rolled-back.

Cultural flourishing

> *What the warming world needs now is art, sweet art. We can register what is happening with satellites and scientific instruments, but can we register it in our imaginations, the most sensitive of all our devices?*
>
> Bill McKibben, American environmentalist

New ideas tend to only take root deeply when they capture the popular imagination through expression in literature, song, drama, and art. In clubs, pubs and community centres across the UK, there are signs that a cultural transition is already underway, and storytelling is at its heart.

Stories help us make sense of the world and prepare us to adapt to change. London is now awash with a revival in storytelling events – the National Theatre put on its first adult storytelling event in 2009, a monthly event at the Canal Café Theatre in London's Little Venice, encourages people to tell their own tales by drawing on personal

Box 3. The Transition Network Movement and other examples of reskilling and localisation

The most striking example of localisation in the UK is the growth of the Transition Network, one of the fastest-growing social movements in the world. The aim is to reduce oil dependence, and lower the ecological impact of local and regional economies by localising the way they meet their needs for food, energy and other goods and services in just the way we have described should happen following the Great Revaluing.

Almost exactly a century after the first garden city was begun on land in Letchworth, north of London, the first transition town initiative was set up in Kinsale, County Cork, Ireland. There are now over 100 transition towns in the UK and many more internationally, with hundreds more under consideration. The South West is trying to become a transition 'region' and Wales is considering being a transition country.

The first stage is to produce an Energy Descent Plan. By 2021, the Kinsale Transition Initiative plans to source the majority of its energy from sources within a ten-mile radius. Its distributed energy system will include wind, biomass from short-rotation coppice, and anaerobic digesters for a combined heat and power plant, and solar.

Similar initiatives in Europe include the community of Vauban, near Freiburg in Germany. It claims to have achieved CO_2 emissions reductions of between 80 and 90 per cent, due to good building design, renewable energy, walking and cycling orientated transport infrastructure, car sharing and good water management.[79]

The dramatic renaissance of urban gardening and horticulture in the UK is also evidence of an unplanned cultural and economic self-medication. In the face of rising food and fuel prices, and the alienation of spending six months of your life shopping in a supermarket, it's clearly meeting a need.

experience. Around the world, initiatives like the Transition Town Movement are spreading through towns and communities each with their own storytelling projects at their hearts.

Since 2001, Cape Farewell has taken scientists and artists ranging from Jarvis Cocker to Anthony Gormley on long journeys to a cold places on the front line of climate change. Seven arctic expeditions have inspired a growing body of artwork and stories. On a gentler level, arts-ecology group Common Ground works to reintroduce us to the places we live in through a campaign for local distinctiveness. In 1990, Common Ground invented Apple Day as a way of celebrating and demonstrating that variety and richness matter to a local area. By 2000, the festival was celebrated in over 600 locations around the UK, and is now a regular event in many towns and villages.

There are also signs of more subliminal cultural change. Designed during the Second World War for use in the event of enemy evasion, it has been hard to avoid the distinctive red and white 'Keep Calm and Carry On' poster in 2009. It can be found from the Prime Minister's strategy unit at No. 10 to the Emergency Planning Office at Nottingham council. This and a new popular interest in the conservation measures taken during the Second World War, point to a kind of unconscious cultural effort to prepare for a greater degree of adaptation.

During the Second World War, the UK Government employed the creative community to encourage changes in consumption not possible by rationing alone. From exhortations to 'dig for victory' to campaigns to 'holiday at home' and the festivities that made it enticing, the success of the home front was rooted in our collective creative capacity. Convinced that creativity could help lead America out of depression, President Roosevelt pumped millions of dollars into the arts. Drawing inspiration from the Mexican mural movement of the 1920s, the New Deal put socially conscious murals in public buildings across the country. An ambitious oral history project documented national history including the first-person accounts of former slaves. The Federal Music Project employed around 16,000 musicians at its peak, who gave an estimated 5,000 performances to audiences of around three million people each week. In 1939, around132,000 children and adults in 27 states received music lessons every week.[80]

In March this year in the UK, artists and writers gathered at 11 Downing Street inspired by an article written by *New Stateman's* former political editor, Martin Bright that drew inspiration from Roosevelt's new deal for the arts. Although still to take shape, the New Deal of the Mind indicates that there may well yet be the political will to support the cultural shift needed to power the Great Transition.

To do so, the Great Transition needs to inspire our creative talents, be reflected in the mass media, and put culture back at the heart of productive activity. Most of all, it must embrace a great democratisation of the arts, turning us from passive spectators in our own lives back into to active participants in it.

Feeding ourselves

The Great Transition will involve a shift to much more local food production. Partly, this should happen naturally as repricing will see the cost of much of the non-local food we buy today from supermarkets significantly rising in price as social and environmental costs are factored in.

The power of price changes to rapidly shift people from food consumers to producers was made clear in 2008 when the price of commodities, oil and food more generally rose very rapidly (bread rose by 20 per cent in a year) leading to a huge growth in the sale of bread makers and chicken coops.[81] In 2009, more than a quarter of us are growing some of our own fruit and vegetables, with a further one in 10 considering that this will be the year they start. Sales of flowers have declined, as sales of vegetable seeds have risen by 128 per cent from last year and ready-to-plant vegetables by 40 per cent. Sales of greenhouses are up by 157 per cent.

However, the scale of the shift towards greater food self-sufficiency will require the public sector to play a leading role, building on and accelerating these trends and encouraging the exchange of skills between people. Following the Great Redistribution, lots of privately owned land will become available to national and local government and could be put into local community ownership and dedicated to the creation of public allotments. The demand is definitely there: in some areas of the UK, waiting lists can stretch to over 10 years and there are now more than 100,000 people waiting to grab a small plot of land and try their hand at growing for themselves.

But while the demand is there, and prices will strongly encourage local production, we will need to relearn skills that have been lost to make this a reality across the country. The national and local public sectors will need to facilitate a huge food-growing training scheme, reinvigorating agricultural and horticultural training and extension services that have declined precipitously in recent years.[82] Funding would need to be expanded for the local community groups that are already dedicating time to local food growing (Box 4), and consumption patterns, too, would need to shift, so that the demand for locally produced seasonable produce met its supply. Such a coordinated approach would enable the type of food-growing revolution that occurred in Cuba after it experienced the shock of losing access to the cheap Soviet oil that fuelled its agriculture. It can be done, but it won't just happen.

Box 4. Growing your own food – examples of community activity

In the London Borough of Hackney, an urban social enterprise called Growing Communities runs an organic food box scheme, a farmers market, and helps people to 'make, bake, grow or pick good food' throughout the year (the fashion for food foraging is on the rise again more generally). It celebrates the annual harvest period with the Good Food Swap, when people exchange recipes, tips and food.

In Sheffield, the Abundance project brings together a team of volunteers who map and harvest seasonal local fruit like apples, pear and plums. They redistribute the surplus to community cafés, nurseries and individuals. Tonnes of fruit gets juiced and made into jams, pickles and preserves. Lessons are given to local school children. They also run seasonal planting and pruning workshops, and published a guide to urban food harvesting.

People will have more time to grow their own food with the reduced working week and less time spent travelling to and from work. Food growing, harvesting and distribution would again become a collective activity for many people who will share community allotments, strengthening social networks, the core economy and well-being. In Cuba, people relearned how to grow their own food together, a key factor in the country's successful transition to food self-sufficiency, particularly the skill-sharing between the old and the young.[83]

The food localisation and reskilling process will also further reduce inequality as a higher proportion of processed, unhealthy food products (which have greater environmental costs) are currently consumed by poorer citizens. Chronic health problems such as obesity and depression will be greatly alleviated as people spend much more time outside taking physical exercise through food growing. Schemes such as Growing Communities in the deprived housing estate of Hackney in East London will become commonplace.

Creating and conserving local energy

Much as with food, the Great Revaluing will make renewable energy much more affordable. Currently renewable energy like wind or hydro-power is not seen as very efficient because its price does not reflect its environmental benefits and because of the high interest rate payments required to meet the upfront capital costs. Both issues will need to be remedied in the Great Transition: the Great Revaluing will incorporate environmental costs in to energy pricing and the Great Economic Irrigation will enable public and regional banks to create interest-free loans for green renewable infrastructure building.

More energy will also be produced at local and regional levels as the social and local economic benefits of a decentralised energy system will also be incorporated into price. Research suggests that small-scale decentralised renewable energy schemes can achieve the UK's energy targets in the most socially, economically and environmentally effective way.[84] This type of energy system also has the potential to overcome the nimby (not in my backyard) attitude that is proving to be such a block on planning for local renewable energy schemes in the UK. The combination of feed-in tariffs and a decentralised energy grid will give people a direct incentive to create their own energy.

In addition, the emergence of co-operative community renewable energy schemes will give people a secure and reliable place in which to invest their savings. People will have a genuine economic stake in creating and investing in energy production. Already, co-operatively owned renewable energy schemes such as Energy4All wind farms have demonstrated the potential for sustainable investment in community renewable energy (Case Study 4).

Every region of the country with serious renewable energy potential (wind, solar, hydro) would develop vocational colleges and universities specialising in renewable energy engineering and production, creating green jobs to crew the Great Transition.

Large-scale, targeted public investments will also go into home insulation, transforming the UK's housing stock from one of the least energy efficient in Europe to Scandinavian standards. Currently, over 27 per cent of the UK's total carbon emissions are the result of the UK's inefficient domestic building stock.[86] Again publicly funded training and education schemes will be required to create tens of thousands of green collar workers. Councils such as Kirklees in Northern England are already showing the way.

Case Study 4. Energy4All Co-operative wind farms

Energy4All wind farms are owned or co-owned by local cooperatives.[85] Share interest is returned annually to members at a reasonable rate of return (around 5–7 per cent) and excess profits are used to support local initiatives like reductions in fuel poverty or conservation trusts. Energy4all now aims to broaden this model by establishing regional investment co-ops which offer democratic, ethical, investments to local people and community friendly capital to renewable energy developers.

Energy4all was created in 2002 due to daily enquiries received by Baywind Co-operative from people looking to replicate the success of Baywind, the UK's first community-owned wind farm. Baywind Co-op has generated enough green electricity to power 1,300 homes a year whilst paying an attractive return to its 1,350 members (averaging 7 per cent per annum), and supporting local initiatives. Owning a wind farm increases awareness of and involvement in renewable energy developments, maximises financial returns from local resources, and mobilises environmental concern.

Getting things and getting things done

As the Internet grows as a means of exchange, it will increasingly liberate people from the formal market and barter and swapping will complement and, in some cases replace, consumption of new and increasingly expensive goods. Already, people sell their own houses and organise holidays, often, without money changing hands in the latter case. But we can also often get tools and fancies for free as people offer up things they no longer need.

The Great Reskilling touches upon many of the kinds of activities that are key to enhancing human well-being. These are the things that, unlike consumerism, really do make a difference to our experience of life, once basic needs have been met. Many of which have been squeezed out in an over-marketised and over-commodified society. **nef**'s research identifies learning, connecting, taking notice, being active and giving. The initiatives outlined above are rich in all of these.

Case Study 5. Kirklees Warm Zone

Kirklees Warm Zone offers energy efficiency measures free of charge to all households. The scheme receives £9 million funding from Kirklees Council, with £11 million CERT funding from Scottish Power. It is area-based, tackling council wards one or two at a time. The key features are as follows:

- Area-based delivery, using contact with community and voluntary groups.
- Intensive marketing campaign and letters to households.
- Trained assessors make door-to-door visits to check insulation status, and also refer people for other services.
- The contractor surveys homes and installs mineral-fibre insulation in lofts and cavity walls.

The average cost per home is £224. So far, 125,000 households have been visited, 88,000 homes assessed and 74,000 referred for insulation surveys. By May 2009, 30,000 had received insulation measures. The scheme is currently installing 500 measures per week. It has also generated 48,000 referrals for measures such as benefits advice, water saving and fire safety.

This scheme has avoided emissions of about 28,000 tonnes of CO_2 per year so far. This will reach about 40,000 tonnes a year by the end of the project. It has created 103 full-time equivalent jobs and developed the skills missing locally to tackle energy efficiency.

Box 5. Online exchange

Skillswap, for example, is a website – www.swapaskill.com – that does exactly what it appears to do. *Instructables.com* is a thriving online community where members exchange instructions on how to produce home-made alternatives to consumer goods. The site has instructions on making everything from chairs, to soap, to solar heating panels.

The Freecycle phenomena is a network now made up of 4,792 groups with 7,194,872 members globally. It describes itself as 'a grassroots and entirely nonprofit movement of people who are giving (and getting) stuff for free in their own towns and thus keeping good stuff out of landfills'. They also have an online café for people to share information, ideas and requests for help.

The Great Economic Irrigation

In the previous two chapters we sketched out the process of localisation in the Great Transition and the reskilling needed to make this reality. We also put this in a regional, national and international context, to emphasise that we do not envisage closed self-sufficient local economies, but rather thriving and vibrant neighbourhoods that exist as part of wider regions and a broader national society, itself part of an interdependent world.

In the Great Transition, finance is fundamental to this. Finance *irrigates* local economies, facilitating sustainable economic activity and underpinning community life. Nationally, we need to finance a renewable energy and sustainable transport infrastructure, as well as private sector activities that extend beyond the local level, including to international trade. In the public sector, tax and spending is needed at both local and national levels to achieve shared, democratically determined goals and to support the provision of national and local public services.

Public financial irrigation

Tax

Nobody likes paying tax. But if we want good local and national public services, a stable economy and well-resourced public goods, these need to be properly funded. This raises two questions: first, what taxes should be raised and spent nationally versus locally; second, what are the most appropriate sources of these revenues?

Whether tax is raised national or locally, the Great Transition would see a major shift from taxing good things like investment and employment to bad things such as speculation and unsustainable consumption.

Given the roles of the national public sector we have described, key functions of the national tax-and-spending system are to maintain broad economic stability and prevent the re-emergence of wide regional inequalities. Three principles of national tax and spending policy are based on these goals in the Great Transition.

First, while income tax for most people would be abolished with the minimum threshold set at the median wage, *income inequality* would be kept within reasonable bounds through the levying of an income tax that rises progressively thereafter.

Second, *asset inequality* at the individual level would be primarily addressed through the inheritance-tax-funded Citizens' Endowment described in the Great Redistribution, but would then be held in check through a progressive capital gains tax, complemented with a national land value tax[87] to create a regional inequality reducing mechanism: if some areas grew more prosperous than others, this would be reflected in higher land values, generating higher tax revenues. Land in less prosperous areas, in contrast, would have lower values and so attract lower tax rates, providing incentives for investment and economic regeneration.

Third, while the tax measures targeting asset inequality would also be positive from general *economic stability* through their dampening effect on speculation in financial and property markets, this would be complemented with transaction taxes levied at a low rate on all financial transactions – including on international currency transaction through the introduction of a Tobin tax. Such taxes would discourage short-term speculation and high-frequency trading but not long-term investment, as the impact of the tax would be negligible in any longer-term investment, but would become progressively larger as the number of transactions rose.

As well as maintaining economic stability and broad levels of equality, the final role of the national tax system would be to discourage the production of environmental 'bads' and encourage the production of 'goods' through variable consumption taxes levied on goods produced outside of local areas. Phased-in in line with income redistribution to address the regressive nature of consumption taxes, the full environmental costs would be reflected in the tax rate levied, reducing demand for goods in direct proportion to the real costs associated with their production.

Environmental *benefits* would also be reflected, to the extent that consumption taxes could even turn negative and become subsidies, reducing the market price of goods with very high environmental benefits and increasing the demand for them in proportion to their ability to build real value.

The combined effect of these national-level tax measures would be to prevent the formation and growth of speculative bubbles in financial and property markets, thus ensuring broad economic stability, to maintain reasonable levels of individual and regional equality in terms of both incomes and assets, and to steer consumption away from things that are environmentally damaging towards those that produce real environmental value.

In the Great Transition, these goals are best pursued at national level for three reasons. First, national economic stability can, by definition, only be monitored and maintained at the national level. Second, a danger in a devolved and decentralised economy of the type set out in this report is that local regions start to diverge in terms of prosperity, and without any national level mechanisms to perform a rebalancing role, this divergence becomes ever more entrenched and pronounced over time. Third, the environmental problems we face are common to us all: things that degrade the environment in one part of the country do no differently in any other.

And spending

What about national level spending? Well, the obvious use to which progressive income tax should be put is in providing a national level safety net, which guarantees a minimum income to those unable to find work, as well as in providing a state pension for older people.

Revenues from capital gains and transaction taxes could be targeted towards providing national level public services, such as specialist larger hospitals and tertiary education institutions, which would require a certain economy of scale and so would be best funded and provided at national level.

The provision of national level public goods could then be funded through the proceeds from the land value tax, facilitating the development and maintenance of a national sustainable energy network to provide back up for local energy production, based on larger-scale facilities, such as offshore wind-farms, for example. Similarly, national transport networks of high-speed rail links integrated into regional and local rail systems and the national road network and could be part-funded and maintained in a similar way.

The purpose of national public spending in the Great Transition would therefore be to partly fund and maintain the things we need but which are best provided at a scale beyond that possible locally, and to provide a basic national income for citizens unable to work as well as a state pension for all.

As should be clear, many issues of taxation and spending are increasingly decided at the local level in the Great Transition. Granting genuinely democratic local government powers to tax and spend – as well as the ability to borrow for local investment – is a vital part of the reinvigoration of local public engagement and participation.

Local authorities have become increasingly marginalised and distant from the lives of people living in local communities. What steps would be taken to reverse this in the Transition? First, local authorities would be freed from the straightjackets imposed by central government and able to borrow to invest in their areas through the issuance of local – or municipal – bonds. Initially focusing on regenerating and retrofitting the local housing stock, it would become possible for local government to provide ongoing investment to maintain a sustainable and stable mix of social, private and rented housing, tailored to the needs of different areas.

Council taxes would become Local Services Taxes and replace income tax for most as their primary means of contributing to the provision of good public services in their areas. Much of this would be dedicated to funding local schools, local preventative health services and community hospitals to treat all but the most severe conditions.[88] Services would become increasingly responsive to democratically expressed local opinion, with a reinvigorated sense of civic pride inspiring much greater participation through volunteering in nurseries, schools, hospitals and care centres for the elderly.

While sustainable locally produced goods would be generally exempt from environmental consumption taxes levied at national level, thereby providing an economic incentive for local production, variable consumption taxes would also be applied locally. Rather than discouraging environmental 'bads' and encouraging 'goods', here the focus is on the social. Reflecting the fact that different communities have different priorities, we would see real participation of local people in priority setting: which social 'goods' do they want to encourage and which discourage? Building the results into local consumption taxes would reflect these priorities in prices and so incentives, but also provide a revenue stream to augment Local Services Taxes in funding services and other local priorities.

After the Great Transition much energy will be produced locally. The introduction of feed-in tariffs[89] at an early stage is crucial to kick-starting this process, but the result would be a far more decentralised, self-sufficient and so resilient energy infrastructure. Connected to and so backed by the national network, this relationship will be mirrored in the transport sector, with local branch networks connected to a high-speed rail system. Many towns would also choose to become pedestrianised, re-establishing affordable trams and other forms of sustainable travel to replace the need for motorised transport in local areas.

Private financial irrigation

In the Great Transition, the perceived divide between the public and private would become far less pronounced. The financial crisis has made it obvious that this was never so clear-cut as had been thought anyway, and the need for a fundamental restructuring of the national banking system – an essential precondition for the transition set out in this report – is clearer than ever. Instead of going back to business as usual we need to build an ecology of financial institutions, established to provide funds to achieve national and local priorities: to irrigate the Great Transition.

Big irrigation

At national level, a Green Industrial Bank would be created to channel public and private funds to build countrywide sustainable energy and transport infrastructures. Rather than relying on private banks as the sole source of money creation, much of this function should be taken back under public control for public benefit, with the Government determining the level of public money that is created annually to maintain a stable money supply. As we have seen with the programme of 'quantitative easing'. the Government can create money at will. Historically, however, it has largely outsourced this job to private banks to create credit, varying interest rates to alter the incentives for banks to do and for borrowers to want to borrow.

The current crisis has made it very clear that this certainly does not guarantee that created credit is put to productive use. Given the huge capital investment required to fund sustainable infrastructure in the Great Transition, taking a proportion of the responsibility for money creation back into the public sector is essential. It would enable large-scale direct public investment to achieve the rapid progress that an uncoordinated market system cannot.

Social challenges are also immense, and addressing these should form a complementary and central role of the new industrial policy. Disproportionately focusing investment in deprived areas, the policy would become a highly effective tool of economic and social regeneration.

As well as the Green Industrial Bank, a national Post Bank would be built out of the existing Post Office system. The Bank would be charged with maintaining a public benefit component in the provision of finance, whether to reduce and eliminate financial exclusion, or to finance small businesses in local communities. The high levels of trust and affection with which the Post Office is held would enable it to rapidly grow its depositor base with new savings, and its extensive branch network would help it to work closely with local businesses and entrepreneurs to provide finance on fair terms.

An existing national institution – perhaps Northern Rock – could be transformed in to a new national Housing Bank, offering people the opportunity to transfer a portion of their mortgage debt into equity and paying social rent on the balance. As well as addressing the millstone of unsustainable personal debt, this would lead to a more balanced tenure distribution across the UK and encourage institutional investment in rented housing, improving the quality of the private rented sector and giving pension funds, for example, an attractive long-term and in many cases local investment opportunity.

The Housing Bank could purchase equity stakes at a reduced market rate, reflecting the future decline in the value of housing because of the introduction of higher inheritance taxes, as well as the puncturing of the credit-fuelled housing bubble. As they would have to book a loss on these sales, other private banks that owned these assets would therefore share the burden of reducing debts, just as they shared in the profits that were made in supplying them. The impact of reducing debt to sustainable levels would be shared by individuals, the public sector and the private sector.

An ecology of irrigation

But the new ecology of finance needs to look beyond crisis management and be varied in scale as well as ownership, form and function. Recognising the importance of financing different activities with institutions dedicated to specific purposes, retail banks should be separated from investment banks. The long-standing tradition of mutual ownership in the UK could finally achieve its potential with many institutions becoming mutually owned, made possible if central government uses its equity stakes acquired in 2008 to hand back ownership to citizens.

At local level, this would see a flowering of local, mutually owned financial institutions, providing a safe home for local savings, but also a source of local investment funds to complement those of the Post Bank.

Privately owned banks complete the ecosystem, with some focusing on particular sectors such as agriculture or providing wholesale banking services at national level, while others provide a wider range of financial services locally.

What these banks would have in common would be new requirements on credit creation. In the Great Transition, public money creation would be a vital tool of the Green Industrial Policy at national level. For non-publicly owned banks, the Government would reintroduce 'fractional reserve requirements'[90] as an active tool of economic policy, with the default level being set at 100 per cent, making credit creation impossible. In practice, however, this percentage would vary to allow credit to be created, but only under strict conditions.

The fractional reserve requirements facing banks would become a direct function of the ability of the investment being financed by the loan to create social and environmental value. As we saw in previous sections, the consumption tax system would be applied differentially to different products to reduce the price of goods in proportion to their social and environmental benefits but to increase prices in relation to costs. This same measurement system could be applied to the financial sector, such that investments that would generate the most value in these areas cause reserve requirements to fall to their lowest possible level of 10 per cent. Where investments would lead to social and environmental costs on a net basis, in contrast, fractional reserve requirements remain at 100 per cent, so that no credit could be created.

As the net social and environmental benefits of projects rise, so fractional reserve requirements fall, ensuring that credit is only created for net positive investments, and the more positive these are the more credit would be available. Keeping interest rates in place makes sense here. Many have criticised the role of interest in driving unsustainable growth and there is much in these arguments. However, when profitability is linked to sustainability and real value, having to earn sufficient profits to service a loan becomes a positive thing: the more social and environmental value is created, the more profitable the investment. Incentivising firms to maximise this value in order to pay interest on their loan reverses today's outcomes, driving sustainability not undermining it.

The result would be a highly effective combination of direct public investment in large sustainable infrastructure projects, and a race to the top in the private sector as reformed financial institutions, at both national and local levels, sought out investment projects that would generate the most social and environmental value. The fact that the money supply could only increase as real value was created would keep supply and demand in balance, preventing inflationary pressures emerging.

Financial institutions in the Great Transition would function within a stable regulatory framework that also encompassed the capital markets. As before the Transition, large liquid bond and equity markets would exist, but would be greatly changed in their approach and impact. Once company profitability becomes directly linked to social and environmental value, share prices for listed companies would reflect this.

The composition of the FTSE 100 would change rapidly in some instances, as ethical and environmentally sustainable companies replaced unsustainable incumbents which were unable or unwilling to change their business practices and so saw their market shares and profitability plummet. In other cases, large companies would change direction and focus on building sustainable social value, while for a few it would be a continuation of their existing good practice, but now they would be rewarded for it and encouraged to go further, rather than facing a financial penalty as had been the case.

The introduction of financial transaction taxes would encourage stability in the stock market, which would be further enhanced by the more long-term, sustainable approach taken by companies to their business activities. This would prove a boon for institutional investors such as pension funds, with their inherently long-term perspective. Pension fund members would also be given a far greater say in the investment decisions of their funds. Given the shift from growth to sustainability in the Great Transition, pension funds – and other investors – would look not to capital gains but to acquiring income streams from sustainable businesses, and to holding a genuinely diversified portfolio of assets with a range of local, national and international investments.

One new area focuses on local investments. As local government becomes able to issue bonds to fund local investments in housing and other areas of social priority, local pension funds would emerge to enable people in communities to invest directly in the future of their areas. Over time, local pensions would begin to be paid on the steady income streams these investments produced, diversified with equity holdings in local companies and perhaps linked to local energy production.

Financial irrigation is essential to kick-starting key components of the Great Transition at both national and local level, but that is only the start. The remodelled public and private financial systems would then help to progressively build value, based on local democratically determined priorities in the public sphere, and the dynamism of financial institutions, now harnessed in the pursuit of social and environmental returns rather than short-term commercial returns, in the private sphere.

The Great Interdependence

I do not want my house to be walled in or my windows blocked. I want the cultures of all lands to be blown about the house as freely as possible. But also I refuse to be blown off my feet by any.

Mahatma Gandhi (1869–1948)

The UK is the world's sixth-largest economy and the impact of its high-consuming lifestyle is exported around the globe.

The clearest demonstration comes from looking at the day in a typical calendar year when, in effect, we stop relying on our own natural resources to support ourselves, and start to live off the rest of the world. The moment we begin living beyond our environmental means is what we call our *ecological debt day*. As the UK's total consumption grows, it moves ever earlier in the year. In 1961 it was 9 July, advancing to 14 May in 1981. At current levels of natural resource use in the UK, the average person went into ecological debt on 12 April in 2009. Looking back, if the whole world had wanted to share UK lifestyles in 1961, the Earth would just have managed with its available resources – one planet would have been enough. But today, if the whole world wanted lifestyles like those enjoyed in the UK we would need 2.5 planets.

At a time of increasing global energy insecurity, the UK's dependence on energy imports has been steadily rising since about 1999. In the four years between 2004 and 2008, the UK's energy dependence factor – a ratio that reveals how much we depend on imported energy – increased 5-fold, rising 30 per cent between 2007 and 2008 alone (Figure 5).

The UK's food self-sufficiency appears to be below what it was 40 years ago. Yet, climate change, new human dietary patterns that put more pressure on land and energy, energy prices and shortages, and competition globally between land for food and land for biofuels, have all increased awareness of the vulnerability of the international food chain on which we depend.

No single country can live entirely within its own means, but the world as a whole does and it, too, is living beyond our ecosystems' capacity to regenerate and absorb waste.

Figure 5. UK energy dependence 1998–2006. [91]

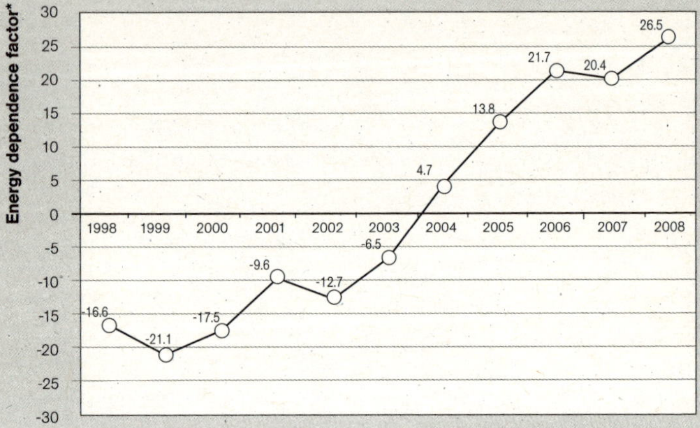

Source: **nef** estimates based on DECC (2009), Digest of UK Energy Statistics (Table 1.1 – 1.3)

** Defined as: Net Imports ÷ Energy Demand x 100, a negative number means that the UK is energy independent because we indigenously produce more than we use. Energy independence does not equate with sustainability.*

Humanity combined now goes into ecological debt on 25 September. Since first going into the environmental red in the 1980s, the over-arching trend has been for the date to move ever earlier in the year.

Overall, it's clear that the UK's patterns of interdependence will have to change if our economy and lifestyles are to become remotely sustainable. In the Great Transition we envisage a global 'deal' to achieve just this. Based on scientific evidence of total global carbon limits and a commitment to global equity and fairness, the UK would progressively reduce its carbon emissions to a level that reflects its size, leaving environmental space for other countries to develop and make their own transitions. Importantly, the UK would also need to help finance this transition. In this report, we assume a transfer of around £200 billion to developing countries over a ten-year period as we progressively move to a long-term path compatible with the UK's fair share of total global emissions.

Scaled up to all developed countries as part of an international deal to avert catastrophic climate change, this would see trillions of pounds transferred to the developing world to fund the transition to sustainable development paths. To put this into context, total global aid flows in 2008 were a little under £80 billion.

Pricing in the environmental costs of transporting goods around the world will inevitably reduce the total level of world trade, but also change its composition significantly. This is happening to some extent anyway. The wisdom of basing development strategies on increasing exports to developed economies has been shaken by the current crisis. A rebalancing of internally and externally focused development is desirable in terms of resilience and the ability of countries to shape their own development trajectories. As in the developed world, more local production and more regional trade are likely to become the norm.

If carefully managed and well-funded, this would see developmental progress accelerate and become less fragile and susceptible to being blown off course by shocks from the international financial system. Less vulnerable development models, coupled with large-scale transfers from richer to poorer countries, could see poverty and global inequality being progressively reduced and the huge dangers to developing countries of irreversible climate change averted.

As is true within the UK, we need to create a more equal world. This certainly requires major redistribution at global level, but also for developing countries to be less buffeted by volatile global markets, and more able to pursue their own priorities with greater autonomy.

This is easy to say of course, but there are very big questions to be answered, which are beyond the scope of this report, before we could begin to move in this direction.

To give just two examples:

1 What will alternative economic development pathways for a post-carbon society mean for patterns of trade, production, consumption, investment and the movement of finance, at a wide range of levels from local to global?

2 Given that any solution to the challenge of climate change must be both global and equitable, how can the North facilitate an alternative development paradigm in the South? And how can the South facilitate a transition to a post-carbon society in the North?

However looked at, the way in which the next international climate change treaty is written will set the context for many of these questions. To lay the foundations for the Great Transition, it will need to have certain characteristics.

It will need to work to stop runaway climate change, in other words to have 'environmental integrity'. And will need to work politically to win the support at least of the major populous countries of the global South, and the wealthy countries like the UK who are home to the major per capita polluters. These countries will need to agree to:

- **Set a formal greenhouse gas atmospheric concentration target likely to keep global average temperature rises below 2ºC.** Paradoxically, whilst many governments including the European Union have accepted the temperature target – they have not agreed on a concentration level likely to meet it. They have, in effect, decided collectively to travel from Kyoto to Copenhagen, but refused to set-off from their base because they cannot agree on what route to take. Now, though, more people are arguing that even a 2°C rise represents a level of warming that is unacceptably dangerous for many of the world's people.

- **Deliver a global, effective and equitable agreement beyond 2012 that deepens reduction targets in industrialised countries**. Wealthy industrialised countries should be setting a legally binding and constantly contracting carbon budgets that plot a course, year by year, towards emissions cuts of well over 80 per cent against 1990 levels. In agreeing the price for full, developing-country participation in a global deal, industrialised countries need to be aware that, to the global South, even equal per capita rights to the global commons of the atmosphere appears as a compromise in the face of historically grossly unequal emissions. A vital part of this process will be to agree binding financial transfers from North to South on a sufficient scale to fund a transition to sustainable development paths.

- **Re-engineer the global economy to allow real poverty reduction in the context of serious carbon constraints.** This means far greater flexibility over economic policy and plugging the leakage from the poor to the rich of the assets and wealth that they do possess. It also means a shift from hoping, with fingers firmly crossed, that a dwindling trickle down of wealth will end poverty as a side-effect of the rich getting richer, towards a more targeted and guaranteed approach of redistribution. This will not be aid, but ecological debt payments by the wealthy that could create, for example, the opportunity for massive investments in appropriate adaptation and renewable energy in poor countries. Sustainable development would also be supported by greater flexibility in the rules governing trade, migration finance and intellectual property.

- **The explicit recognition and protection in international law of environmental or 'climate' refugees, displaced due to push factors resulting from global warming**. Measures to meet this large and growing challenge should operate along the lines of appropriate burden sharing. Financial burdens should be met proportional to both responsibility for the problem and ability to pay. More than that, there will need to be flexibility in immigration policy, protection of displaced people and compensation fund for those affected.

- **New and additional resources made available for adaptation.** In yet another climate change irony, in 2008 the rising price of oil was reported to have wiped out the value of aid to Africa.[92] While the costs of adapting to climate change in the majority world are large and unknown, new resources available to pay for adaptation are, conversely, known and small. Funds for this purpose under the United Nations Framework Convention on Climate Change (UNFCCC) and Kyoto Protocol need to increase in size by several orders of magnitude, in order to match the costs of unavoidable adaptation and pay for clean energy substitution.

- **Technology transfer.** Free technology transfer is also important, especially energy technology, unconstrained by the restrictive regimes governing intellectual property in the global economy.

We are all in this together. No country can overcome the challenges we face alone, but this is not simply a matter of enlightened self-interest. The wealth of richer countries has often been based on the exploitation of poorer countries. The environmental catastrophe that looms is the result of the cumulative emissions that our financial wealth in the developed world has been built upon. There is a clear moral case for compensation, but also an overwhelming need for us to find a new global dispensation, and one that recognises that all who share this world have an equal right to live a meaningful and flourishing life.

We are already interdependent. To create a global Great Transition we must, in many important respects, become more so.

Two big remaining challenges

Six decades ago, Karl Polanyi analysed and set out a vision for addressing the fallout caused by the encroachment of market processes on the fabric of social life. This report has aimed to do something similar but in response to the very specific – and urgent – social, environmental and economic problems we face today.

Achieving the Great Transition will not be easy, but we believe it is possible. To get there, however, we will need to overcome some major challenges. Here we focus on two in particular.

Getting the prices right

We need to rein in the market and create space for non-market activities in the core economy to flourish, but we also need to ensure that market activity drives rather than destroys social and environmental value. This cannot happen unless we get the prices right. Unless prices reflect real social and environmental costs and benefits, profit will always come at the cost of people and the planet. But if harnessed to these ends, the market could allocate scarce resources such that social goals are advanced within environmental limits.

The price mechanism is a powerful force acting on behaviour, with the potential to steer and coordinate economic activity across local and national economies so that these goals are achieved. For this to happen, however, we have to get the prices right.

Measurement is at the heart of this. For prices to reflect social and environmental 'goods' and 'bads', we need to be able to accurately measure these externalities. We know how to do some of this already: many companies now factor in a future carbon price in their investment decisions, knowing that this will ultimately be a real cost to them.

This is, however, a complex process. Where climate change is concerned, price mechanisms can only work effectively once a scientifically arrived at safe level of emissions and concentrations of greenhouse gases has been set. This is why we have argued strongly for a quantitative cap on carbon emissions to set strict parameters within which pricing can be used effectively.

Progress has been made on the pricing of ecosystem services also, although this is less advanced than with carbon. We need to consolidate, build on and rapidly accelerate work in this area with the aim of creating approaches that are widely seen as accurate, but are also manageable so that they can be practically integrated into our day-to-day lives in real economies.

This will be no easy task, but it is not impossible. It is also essential if the Great Transition is to become a reality.

Getting the economics right

> *Put bluntly, the dilemma of growth has us caught between the desire to maintain economic stability and the need to reduce resource use and emissions. This dilemma arises because environmental impacts 'scale with' economic output; the more economic output there is, the greater the environmental impact – all other things being equal.*
>
> Tim Jackson, *Prosperity without growth* .

Of course, as Tim Jackson points out in his recent report for the Sustainable Development Commission, all things are *not* equal.[93] While there is now widespread acceptance of the severity of our environmental position, there is a presumption that growth can be decoupled from environmental impact through more efficient use of resources and through the dematerialisation of economies.

As Jackon argues and as **nef**'s own climate research has shown, there is absolutely no evidence to support this – quite the opposite in fact: the scale of output continues to outstrip efficiency gains and no economies have dematerialised to any meaningful extent or show any signs of doing so. The reasons for this have long been well understood, though largely ignored. The environmental economist Herman Daly put it like this: 'The notion that we can save the "growth forever" paradigm by dematerialising the economy, or "decoupling" it from resources, or substituting information for resources, is fantasy. We can surely eat lower down the food chain, but we cannot eat recipes…'[94]

It looks increasingly inescapable that we need to stop the growth in output and consumption and even to put this process into reverse – at least until an environmentally sustainable level is reached, taking full account of all potential efficiency gains and renewable energy uses.

The problem is that our economies are entirely geared towards maximising growth. Whilst this may not always succeed, there have been no attempts to deliberately engineer a 'steady-state economy' of the kind that would be needed after the Great Transition.

The Canadian economist Peter Victor is one of the only members of the profession to have devoted any time to this subject – which in itself is astonishing. Victor constructed a macroeconomic model of the Canadian economy, where standard macro variables (e.g., savings, investment, output, consumption, public expenditure) combine to produce outputs (e.g., poverty, employment, debt, GDP per capita, $C0_2$ emissions, etc.), with the relationships based on empirical data from the Canadian economy.[95]

Victor then proceeds to develop a number of scenarios by altering the variables associated with growth – such as investment, for example – and looks at the impact on key variables. The first scenario shows why everyone is terrified of even talking about no longer targeting growth: GDP per capita remains flat and CO_2 emissions fall, but unemployment, debt and poverty escalate alarmingly.

The second scenario is more encouraging, however. As we do in the *Great Transition* report, Victor assumes an introduction of measures such as shorter working weeks, more publicly funded investment on infrastructure and the provision of high-quality public goods. Here, CO_2 emissions also fall, but so, too, do unemployment, poverty and debt.

This is an important first step, but no more than that. What Victor has done is to adapt standard macroeconomic models, but it is clear that a new modelling approach is needed. Standard models take no account of the use of finite resources and environmental constraints, and are blind to social outcomes in terms of equity and, of course, human well-being.

To make a reality of the sort of society and economy envisaged in this report, such a model is not a luxury but an essential foundation.

Macroeconomic models are open-ended by nature, with growth being the primary output of interest. Inputs feed in, interact with each other, achieve balance (or equilibrium) and outcomes result. We need to reverse this. That is, to start with the hard outcomes we need: environmental sustainability; equitable social and economic justice; and high levels of human well-being. We then propose to link these to relevant economic determinants within the model (aggregate output, income distribution and working hours, respectively, for example) and to 'reverse engineer' what this would imply for the levels and types of differing inputs.

Solving the problem of economic externalities and the production of a working, full-scale and robust model of the 'new economy' are huge tasks. While we clearly could not do these alone at **nef**, we will continue to contribute what we can to overcoming these major challenges.

Both have to be cracked. We need to get on with the job of doing so.

How to take the first steps

Environmentally, time is running out.

Economically, much of what we have been told was true has been shown to be false, and we are at a once-in-a-century crossroads when a different path must be taken.

Socially, as the corrosive effects of inequality deepen they become harder and more expensive to reverse.

To achieve the Great Transition will require fundamental changes to the way we live our lives, but we believe that the rewards from doing so will be great. These changes are required at individual and government level, and they are inextricably linked: each on their own is necessary but not sufficient, together they make the transition possible.

There are some things we can do straight away, changes that may seem small but when taken by millions together are much more significant. But, we should not pretend that this will be enough. Our lives are shaped by the systems – economic, political, legal and social – that surround us, and it is only by changing these that the Great Transition can be realised. Governments set the 'rules of the game', unless the parameters are altered there is a limit to what we as individuals can do.

But even for government to change they require our support. As the divisions and mistrust within our society intensify it will become ever more difficult to act collectively. Achieving the Great Transition will require us to jump together: to make changes to our own lives, to push the government for radical reform and then to be prepared to see these changes through. We need to be in this for the long haul – to keep our eyes on the bigger picture, the long horizon and the ultimate prize.

Below is a set of recommendations for both governments and individuals. They are far from exhaustive, but we believe that with them we can take the first steps on a necessary and desirable journey to the Great Transition.

What you can do…

… in your own life

- Think about the carbon and ecological footprint of your lifestyle, and take steps to reduce it. Take fewer but longer holidays to reduce carbon emissions associated with travel.
- Cycle or walk whenever you can.
- Start growing your own fruit and vegetables.
- Buy locally produced goods when you can.
- Repair, reduce, recycle and reuse as much as possible.
- Buy only what you need and pay attention to the 'real' cost of goods.
- If you can afford to spend more on environmentally and socially responsible goods, then do so.
- Invest in the energy efficiency of your home. This will save you money in the long run, and you might even be able to take advantage of a grant or subsidy.
- Seek a balance between your work life and home life.
- Get to know your neighbours and become involved in your community.
- Join a group that is campaigning for change, such as a local Transition Town initiative.

… to influence government and others

- Get political. Lobby your MP, the Minister for Climate Change and Energy and Prime Minister to support a global fair deal on climate change.
- Use your vote at every election to support the party with the best policies for achieving a socially just and environmentally sustainable Britain.
- Educate others. Get out, get active and be a positive force for change.

What government can do for…

… the environment

- Agree a global fair deal on climate change with appropriate contraction and convergence targets to avert dangerous climate change, reflecting the UK's 'fair share' of total sustainable carbon emissions.
- Put in place a plan to rapidly decarbonise the economy and hit these targets.
- Challenge the assumption that all value is economic value, and that the interests of business are synonymous with those of society by initiating a 'national debate' about what we value environmentally and socially to kick off 'The Great Revaluing'.
- Instigate a review of taxation to shift the burden of taxation to environmental and social 'bads', based on the issues of concern that emerge from this 'national debate.'
- Start training programmes to develop 'green' skills for the Great Reskilling.

… the economy

- Restructure the financial system so that it better serves the needs of society. Start by creating a Green Investment Bank, Post Bank and Housing Bank and encouraging local mutually-owned co-operatives.
- Use public money to finance the creation of a national green energy and transport infrastructure and provide incentives to kick start local equivalents through under-utilised mechanisms such as local authority bond issues.
- Introduce a windfall tax on the historic surplus profits of the fossil fuel companies, similar to what Norway has done, and use the proceeds to invest in the UK's transition to green energy.
- Link the ability of banks to lend with the creation of social and environmental value to encourage investment flows to enterprises that will build a better tomorrow.

- Introduce a financial transaction tax to discourage the speculation that leads to dangerous economic volatility.

- Regulate the financial system for stability and long-term benefit.

- Progressively introduce requirements for businesses to internalise the negative social and environmental costs of their activities. Start with a mandatory carbon price, and move onto other areas including eco-system resources and community well-being.

- Implement a reduction of working hours and more flexible working arrangements for all employees, to prioritise employee and community well-being through sharing both working and non-working time more equally.

- Ensure that businesses make visible and account for not just financial value, but the creation and destruction of value across a triple bottom line. Put holistic value at the heart of decision-making.

- Introduce regulation to increase employee ownership and their participation within governance and decision-making.

… social policy and social justice

- Devolve real decision making power to democratically accountable local forms of government based on genuine engagement and participation.

- Free these bodies to raise their own finance and spend the revenues on locally determined priorities.

- Reduce a wide range of social ills though cutting inequality by beginning a progressive redistribution of incomes that creates value for individuals and better outcomes for society.

- Investigate the use of carbon reduction measures to help redistribute income to poor households.

- Put in place measures to redistribute wealth and provide all citizens with an 'endowment' on reaching adulthood.

- Put prevention at the heart of public services by investing in initiatives that promote better and more equal life chances, such as quality universal child care from an early age. This will save money in the long run.

- Promote co-production in the design and delivery of public services, as a way of preventing needs arising and increasing well-being, as well as making better use of human assets to meet needs.

- Make public spending decisions on the basis of triple bottom line outcomes including environmental, social and economic criteria, and consider these outcomes over appropriate timescales.

Some of these measures will take time to get going, but others can be started right away. We can all do things in our own lives now. We can encourage others to do the same and lobby government to change the 'rules of the game' to drive real and lasting change.

We should not be put off by the scale of the challenge, but rather take inspiration from what can be achieved if we work together collectively to build a better future. We need to get started.

It is time to start walking as well as talking.

Appendix 1: Climate change targets

Currently, even the most stringent targets are almost guaranteed to push global temperatures 2°C or more above pre-industrial levels (equivalent to a further 1.2°C rise from today. This is because current climate models may underestimate the impact of long-term climate change for a given concentration of greenhouse gases.[96] Research published in 2008 by James Hansen and his colleagues at Columbia University in New York argue that atmospheric concentrations of CO_2 should be stabilised at 350 parts per million (ppm): 'If humanity wishes to preserve a planet similar to that on which civilization developed and to which life on Earth is adapted, paleoclimatic evidence and ongoing climate change suggest that CO_2 will need to be reduced from its current 385 ppm [*now approximately 390 ppm*] to at most 350 ppm.'[97]

It is worth noting, however, this figure excludes other, non-CO_2 greenhouses gases.[98] CO_2 equivalent (CO_2e) is a unit that accounts for other greenhouse gases in the atmosphere are weighted by their 100-year global warming potential.[99] Hansen's reason for focusing on CO_2 is due to its long atmospheric lifetime compared to other greenhouse gases.

However, the relationship between temperature change, concentration of greenhouse gases in the atmosphere and CO_2 emissions is more complex than originally thought. The uncertain relationship between concentrations of CO_2, carbon emissions and temperature change means that researchers are beginning to call for policies to be based on total amount of CO_2 emitted rather than temperatures or concentrations.[100, 101, 102]

A team of researchers published two papers in the journal *Nature* in early 2009 arguing that to reduce the chance of global temperatures exceeding a 2°C temperature threshold, specific caps on carbon emissions need to be set.[103,104] For example Malte Meinshausen from the Potsdam Institute for Climate Impact Research and his colleagues found that to reduce the probability of exceeding 2°C to 25 per cent, cumulative CO_2 emissions between 2000 and 2050 need to be capped at 1000 billion tonnes (Gt) of CO_2 (1,500 Gt CO_2e).[105] To reduce this risk by a further 5 per cent, emissions need to be capped at 890 Gt CO_2 (1,356 Gt CO_2e) or less. Given that between

2000 and 2006, 264 Gt CO_2 were emitted – this means if rates of CO_2 are kept at their current rate of 36.3Gt per year, the total carbon budget would be exhausted by 2024 or 2027 depending on the accepted probability of exceeding 2 °C (20 per cent and 25 per cent respectively). However, the authors also warn that if global greenhouse gas emissions are still more than 25 per cent above 2000 levels in 2020, the probability of exceeding 2°C rises 53–87 per cent. Given that 80 per cent of greenhouse gases are due to the combustion of CO_2, this means limiting use to less than one half of the proven economically recoverable oil, gas and coal reserves.

Using a different methodology, the second paper led by Myles Allen, Head of the Climate Dynamics group at University of Oxford's Atmospheric, Oceanic and Planetary Physics Department yields results that are broadly consistent with Meinshausen.[106] More recent work still from the Met Office Hadley Centre warns of a scenario in which a 4°C rise in temperature by 2060 is possible.

In 2008, cautious calculations by **nef**'s climate change and energy programme suggest that there may be as little as 100 months, starting from August 2008, to stabilise concentrations of greenhouse gases in the atmosphere – before the risk of uncontrollable global warming occurring increases significantly.[107] This has also been supported by the recent research by the Tyndall Centre for Climate Change Research.[108]

Appendix 2: Individual value

The model is based on Layard, Mayraz and Nickell's (2006) study on the rate of diminishing marginal utility of income in over 50 countries between 1972 and 2005. Layard et al. estimate a conservative parameter () of 1.26, which means that the marginal utility of income declines somewhat faster than in proportion to the rise in income. Using a conventional formula of individual utility [U=(y^ (1-)-1)/ (1-) for all <> than 1] and Layard et al.'s parameter, we calculated the marginal utility of an extra £1 for people holding from £1000 to £190,000. A simple ratio of these marginal utilities reflects the added value of an extra £1. For instance, the value of £1 for someone holding £10,000 is 18 times greater than the value of that same £1 to someone holding £100,000.

We then looked at the distribution of income in the UK to estimate the value that would be generated by transferring income from the top to the bottom of the distribution. In particular, we looked at mean net income before housing costs, after taxes and benefits, by decile, and fitted the curve into a continuous one. Since the preferred benchmark was Denmark's Gini coefficient, we reversed the mechanics of calculating the Gini and estimated the total amount that needed to be redistributed in order to arrive at such a figure. More specifically, and considering the UK's current household income distribution, we estimated the total amount that would need to be distributed from the top to the bottom to obtain a more equal (and adequate) distribution. We intentionally decided to preserve the mean income of £24,844. The redistribution was structured to take place in several stages. It is important to point out, however, that the amounts transferred in each stage were chosen arbitrarily, bearing in mind the overall objective of reaching a Gini coefficient similar to Denmark's.

In the first stage, we took income from the top 10 per cent and almost doubled the income of the bottom 1 per cent (from £6,700 to £10,500). We then continued to transfer income such that each percentile received 97.4 per cent of what the previous percentile had received. This was done in order to preserve the continuity and positive slope of the income curve. The change in utility per percentile of the population was also calculated, arriving at the figure of a 4.63 per cent increase in the overall utility (or social welfare). In the second stage, we raised income for the bottom 5 per cent to £11,500 and continued with the same criteria of decreasing the transfer by 4 per cent up to individuals earning £25,992. Total utility (social welfare) at this stage increased by 5.80 per cent. In the third stage, we increased the incomes of the bottom 5 per cent to £12,500, generating an increase of total utility (social welfare) of 6.88 per cent. In the fourth stage, income was increased to £13,900 for the bottom 10 per cent of the population. This figure was chosen as the acceptable minimum wage as it concurs with the findings of recent Joseph Rowntree Foundation's (JRF). This transfer required income to be taken from the top 15 per cent of the population in order to respect the criteria followed in previous stages. Total utility (social welfare), however, increased considerably to 8.13 per cent and the Gini coefficient reached its target of 0.24. A fifth and final stage slowly increased incomes for people between the bottom 5 per cent and 10 per cent, increasing total utility (social welfare) to 8.49 per cent.

Overall, the P90/P10 ratio decreased from 4.40 before the redistribution to 2.73 after the final stage of the process. As mentioned before, the Gini coefficient also decreased from 0.33 to 0.24 and the bottom 5 per cent of the distribution end up earning the minimum income required to obtain a socially acceptable standard of living. Changes in household income distribution curves before and after are shown in the figure below.

Appendix 3: The social value of greater equality

More unequal societies do worse on a range of social measures from health to education and crime. The preventable problems associated with inequality are costly to the state, and make society a less pleasant place for us all. To calculate the social value that could be created by moving to greater equality we used Denmark as our benchmark as this had also been the basis for our redistribution model (see Appendix 1).

We first collected data on the incidence in the UK and Denmark of a range of social problems that have been associated with inequality. Most of the data was drawn from OECD studies. Outcomes were selected on the basis that they had a relationship to inequality, were costly yet preventable and could be evidenced with good quality comparable data in both countries. Using the incidence data, the cost of inequality in each country was then calculated in UK terms. We have also included wider social costs that relate to each of the problems where possible.

The following costs were included in the model:

- Productivity losses from 16–9 year olds being Not in Education, Employment, or Training (NEET) and lost earnings to individuals.
- NHS costs and productivity impacts of obesity.
- Costs to the State and wider social costs of crime.
- Welfare and health costs of teenage births.
- Welfare and health costs of substance misuse and costs to individuals and communities of drug use.
- Costs to the State and wider economy of mental health problems, as well as costs to those who experience mental health problems.

- Costs to the State and families of family breakdown.
- Regeneration costs from attempts to off-set spatial inequality.

In the business-as-usual scenario, the annual costs of addressing preventable social problems associated with inequality were added up to arrive at a cumulative total for the period up to 2050.

In the alternative scenario, it was assumed that outcomes converge over time with Denmark as the UK reduces its levels of inequality to a gini of 0.24. Controlling for population, the cost-savings and wider social value was calculated annually and then added up over the period to 2050 to arrive at a cumulative total.

Calculating the costs of inequality

This section describes in more detail the rationale and method for calculating costs for each social outcome. Unless otherwise stated, data were drawn from the OECD Society at a Glance indicators, using the latest year available.[109]

NEETs

Data comparisons were drawn from 2006; the UK had the highest proportion of NEETs within our sample with 10.9 per cent of young people. Much research has been done on the detrimental long run effect that this has on young people; they will often move in and out of employment for the rest of their lives.[110] Costs have been estimated at around £100,000 per year.[111] These costs would tend to overlap with the other outcomes that made up our composite; therefore we only looked at annual productivity losses, which were not being captured elsewhere. For the individuals for whom being NEET was avoided we included a conservative loss of earnings of £1,714. This amounted to annual cost of £5,365[112] for 10.9 per cent of young people aged 16–19 in that year. Costs for other countries were then calculated in UK terms.

Obsesity

In 2007, the estimated cost to the NHS is 17.4 billion pounds. The indirect costs, including things such as absence from work, is estimated at 15.8 billion pounds. The total social cost of obesity is hence 33.2 billion pounds.[113] These results are specific to England. If you assume the same level of obesity and costs across Wales, Scotland and Northern Ireland, the costs would be, respectively, 3.25 billion, 1.95 billion and 1.1 billion. The total for the whole of the UK would be 39.5 billion.

The measure used for this indicator was percentage of population with a BMI >30 and data were drawn from 2005 and 2006. This accounts for a significant proportion of the overall costs as at 24 per cent it is almost double the average of the other EU countries. The costs per person of obesity are estimated at £2715 and this has been multiplied by 24 per cent of the population in 2006 to arrive at our baseline cost. Unlike some of the other social problems this is also predicted to be on a sharp upward trajectory with studies suggesting it will hit 40 per cent by 2030.[114] Despite these projections, we have taken a very conservative view, with our projections of future costs being based on population increases rather than social trends. Our estimates are therefore likely to significantly underestimate the real costs that lie ahead if these problems are not tackled now.

Crime

Crime is notoriously difficult to measure and compare across countries. For this indicator we have used the OECD crime victimisation study.[115] Although the UK comes second last in our sample, given the disproportionately high rates of incarceration this also underestimates the relative cost of crime. A Home Office study from 2000 estimates the cost of crime at £60 billion a year. Adjusted for inflation, the cost of crime in 2009 is about £78 billion. The figures are specific to England and Wales. With similar interpolation, the number for the whole of UK would be £84.8 billion. We have used this figure to estimate the average cost per crime, which is about £6680. This is consistent with other estimations of the cost of crime.[116]

Teenage births

Although high levels of teenage pregnancy have been associated with poorer outcomes for mother and child in later life, we found that most of these overlapped with other outcomes such as being NEET, and were likely to be heavily rooted in a young woman's experience of poverty and exclusion that is accounted for elsewhere. We therefore only calculated a small proportion of the increase in income support, as well as health and social work costs that would arise from high levels of teenage pregnancy. This amounted to £11,250 per person per year (£10,000 benefits[117] plus £1250 health and social costs[118]).

Substance misuse

The indicator used for this outcome was the annual prevalence of abuse as a percentage of the population aged 15–64 and data was drawn from the European

Monitoring Centre for Drugs and Drug Addiction.[119] The cost per user was estimated at 12,500[120] which translated into a total cost of £77.7 billion. We have also captured the benefits to individuals, families and communities of being free of drugs, and have used a proxy of the amount spent on drugs by a problematic drug user (£16,500) to do so.[121] This represents the value to them of not having chaotic and damaging lifestyles. This arrived at a total cost of £29,000.

Mental health problems

Data for the prevalence of mental health was drawn from the Eurobarometer for 2003.[122,123] This is gathered through face-to-face interviews focusing on 'experiences of one's mental health status in terms of current symptoms of depression or anxiety, sense of psychic distress, sense of positive experience of energy and vitality and awareness of a social support networks.'[124] As this places the UK significantly above European countries at 31.5 per cent, we are assuming that this includes quite low levels of mental health disorder that might not necessarily have a cost attached to them. A King's Fund study has found that there are 8.65 million people in England are dependent on mental health services. Extrapolating for the UK this amounts to 55 per cent of that cohort.[125] Assuming costs hold across the UK, we have calculated a cost for that group for service dependence (2600) arriving at a total annual cost of £26.8b. Incidences of mental health are on the increase and predicted to rise by 14 per cent by 2026.[126] Again, however, our projections are conservatively based on population growth only. We have also incorporated a cost for individuals that suffer from mental health problems, using a proxy of annual counselling costs of £2587. The total cost therefore is £5187.

Family breakdown

Although not necessarily always a negative, this was the most robust indicator for domestic disharmony that was available. Domestic violence would have been more appropriate but data is not routinely gathered to evidence this, and cultural differences in reporting across countries mean that what is available is unreliable. Family breakdown has been estimated to cost anything between £4 billion and £15 billion a year in benefits, legal and welfare costs.[127] We have taken a mid-way point of £10 billion, and calculated the cost per breakdown to be in region of £1543. As the cost is greatest to the individuals and families experiencing breakdown, we have included a proxy of average divorce costs (£13,000). The total cost was therefore £14,543.

Spatial inequality

The UK spends on average £5.5 billion a year to off-set the costs of area-based deprivation through regeneration programmes.[128] As we don't have a measure of spatial inequality across Europe, we have used income inequality as a proxy. The assumption is that this spending would be avoided if these inequalities were not allowed to arise in the first place. This is a conservative estimate, as it is only a part of the costs incurred by local and central government in attempts to regenerate deprived areas. As the UK has the highest level of inequality, for each of the other countries we assumed they would spend a reduced percentage of that depending on their level of inequality. This potentially underestimates the true cost of spatial inequality, as it doesn't take account of people's experiences of living there, which it arguably should.

Appendix 4: Climate change scenarios

Emission trajectories

The emissions trajectories used in this analysis are based on two sources. The reference scenario (business as usual) provides a baseline projection of energy usage, or carbon emissions, within the present stated policy framework. It is taken from the World Energy Outlook (WEO) Update 2008.[129] It considers the effect of government policies up until mid-2008, but not new ones and is based on annual GDP growth of 2 per cent. As such, for the reference scenario, global CO_2 concentrations are expected to double by 2100, reaching 700ppm (CO_2 only) and 1000ppm (CO_2e). The emissions trajectory for the UK is estimated from the WEO reference scenario regional CO_2 growth rates for the EU-27.

The alternative scenario has been sourced from a recent paper from the Tyndall Centre for Climate Change Research, Manchester.[130] The emissions trajectory is based on an equitable apportionment regime ('Contraction & Convergence') of a global carbon budget required to keep concentrations of CO_2 below 450ppmv. While there were three emissions trajectories, in our analysis we used the emissions trajectory with the earliest peak year of 2012 and therefore the lowest cumulative emissions. This would commit the UK to a carbon budget of 16.9 $GtCO_2$. To meet this target, this would require an average annual emission reduction of 9 per cent per annum. It is noteworthy that this is 100ppmv over the recommended long-term atmospheric concentration of CO_2 by Hansen *et al* (2009).[131] As such, we note that the findings drawn from our alternative scenario analysis are conservative and emissions will have to fall more rapidly, exceeding 9 per cent per annum.

In both scenarios, the population trend to 2050 is taken from UN projections.[132]

Estimating GDP in alternative scenario

To assess the level of economic output compatible with the alternative emissions scenario we first examined the reduction in emissions that could be obtained from increased energy efficiency and from reductions in the carbon intensity of energy. For the former, we used Tyndall Centre estimates of the lowest possible energy use in different sectors of the economy.[133] This was one of the only projections of energy intensity reductions by industry to 2050 available. In aggregate, they state that energy intensity could fall by a factor 9.5 to 2050. However, this figure comes with a strong caveat: the Tyndall Centre describe it as the lowest conceivable energy use and the result of a thought exercise, rather than in-depth research. For the energy intensity of carbon, we used the WEO estimate of a 2.8-fold reduction to 2050. Combined, the fall in energy intensity and reduction in carbon intensity of energy is not sufficient to provide the 36-fold reduction in emissions. Output, or GDP, has to fall by a factor of 1.35 from £1.384 trillion to £1.024 trillion in 2050 to meet the emission targets. It must be stressed that GDP will likely have to fall by more than this as the modelling of the likely energy intensity reductions are highly optimistic.

Climate damage cost

The measures of climate damage are based on estimates of social cost of carbon emissions from the Stern Review. It is worth noting that these estimates are today regarded as low in light of more recent scientific findings.[134] The damage caused by each tonne of carbon depends on the final greenhouse gas concentration in the atmosphere. Consequently, we used different damage estimates for the different climate scenarios based on assumptions about the development of global emissions. The WEO reference scenario used the Stern Review's business-as-usual estimate of £57 of damage per tonne of carbon in 2010, while the Climate Act targets scenario used a 550ppm scenario with a £20.1 cost. The alternative scenario assumed a global climate deal and used the 450ppm cost of £16.8. All costs were assumed to grow 2.4 per cent per year, corresponding to the technical notes of the Stern Review.[135]

Abatement cost

To establish abatement costs, first we estimated emission reductions that would take place without any additional investment – i.e. changes to energy/carbon intensity that would occur autonomously. This was based on research by consultancy firm McKinsey. Their research assumed an annual baseline reduction of 2.2 per cent.[136] Over the time period of analysis, however, evidence suggests this is over-optimistic.[137] Given this, the figure was halved to 1.1 per cent. To establish the cost of achieving further emission reductions, we used estimates in the 2007 Energy White Paper which states that an abatement of 226 MCO_2 would cost £29 billion in 2020.[138] The abatement cost of emission reductions beyond the autonomous reductions are assumed to correspond to this cost, in proportion to the required amount of abatement.

The reference scenario that corresponds to Climate Act targets was modelled in the Energy White Paper 2007. Their estimates were that the abatement costs required to meet these targets would be 1.65 per cent of GDP in 2020 and 0.9 per cent of GDP in 2050. In their modelling GDP was assumed to grow annually by 2 per cent, which allowed these shares to be translated into absolute costs.[139] The WEO scenario assumed the same abatement costs, but for a smaller amount of mitigation, using the emissions of 2010 as a baseline.

International trading

International trading of carbon permits was introduced to the model by estimating a maximum level of abatement that increased over time. All the required reductions that went over this level of abatement were handled by international carbon trading. The maximum level of abatement started with 15 Mt CO_2 in 2012 and increased in an exponential manner to 79 tonnes in 2019, after which no international carbon trading was required. The price of carbon on international markets was assumed to match the marginal social cost of emissions.

Endnotes

1 Polanyi K (1944) *The great transformation: the political and economic origins of our times* (Boston, MA: Beacon Press).

2 For a robust defence of economic growth, and its attendant personal aspirations, in moral terms see Friedman B (2005) *The moral consequences of economic growth* (New York: Alfred A Knopf).

3 Wilson TD and Gilbert DT (2005) 'Affective forecasting: Knowing what to want' *Current Directions in Psychological Science* **14**: 131–134.

4 For some recent reviews of this evidence, see Kasser T (2006) 'Materialism and its alternatives' in Csikszentmihalyi M and Csikszentmihalyi IS (eds) *A life worth living: contributions to positive psychology* (Oxford: Oxford University Press); Hnilica K (2005) 'The influences of materialistic value orientation on life satisfaction' *Ceskoslovenska Psychologie* **49:** 385–398; Nickerson C, Schwarz N, Diener E and Kahneman D (2003) 'Zeroing in on the dark side of the American dream: a closer look at the negative consequences of the goal for financial success' *Psychological Science* **14:** 531–536.

5 Solomon S, Qin D, Manning M, Chen Z, Marquis M, Averyt K, Tignor M and Miller H (2007) *Climate change 2007: The physical science basis. Contribution of Working Group I to the Fourth Assessment Report of the Intergovernmental Panel on Climate Change* (Cambridge: Oxford University Press).

6 Millennium Ecosystem Assessment (2005) *Ecosystems and Human Well-being: Synthesis* (Washington, DC: Island Press).

7 Hubbert M (1956) Nuclear energy and the fossil fuels, *Drilling and Production Practice* 7–25.

8 IPCC (2007) *Climate change 2007: Synthesis Report. Contribution of Working Group I, II and III to the Fourth Assessment Report of the Intergovernmental Panel on Climate Change* (Geneva, Switzerland: IPCC).

9 Speech at the International Scientific Conference on Climate Change in Copenhagen, 10–12 March.

10 Hansen J (2005) 'A slippery slope: how much global warming constitutes 'dangerous anthropogenic interference'?' *Climatic Change* **68**: 269–279.

11 Assessing the likely impact of different atmospheric concentrations is difficult, and is explored further in the appendix on climate change targets.

12 Johnson V and Simms A (2008) *100 Months: Technical note* (London: **nef**). Available at: http://www.onehundredmonths.org

13 Anderson K and Bows A (2008) Reframing the climate change challenge in light of post-2000 emission trends *Philosophical Transactions of the Royal Society, A* **366:** 3863–3882.

14 Global Carbon Project (2008) Carbon budget and trends 2007. Available at: www.globalcarbonproject.org [26 September 2008].

15 Klare M (2004) *Blood and oil: The dangers and consequences of America's growing dependency on imported petroleum* (New York, US: Metropolitan Books).

16 IEA (2008) 'Despite slowing oil demand, IEA see continued market tightness over the medium term,' available at http://www.iea.org/Textbase/press/pressdetail.asp?PRESS_REL_ID=267

17 Simms A (2008) *Nine meals from anarchy: Oil dependence, climate change and the transition to resilience* (London: **nef**).

18 DTI (2003) A ready reckoner looking at the effect on fuel poverty of various price and income changes (London: DTI, now DBERR).

19 National Audit Office (2006) *Enhancing Urban Green Space. Report by the Comptroller and Auditor General, HC 935 Session 2005-06* (London: The Stationery Office);

20 Hartig T, Evans G, Jamner L, Davis D, and Garling T (2003) 'Tracking restoration in natural and urban field settings' *Journal of Environmental Psychology* **23**: 109–123.

21 United Nations Commodity Trade Statistics Database (2005). Available at: http://comtrade.un.org

22 Adapted from Millennium Ecosystem Assessment (2005) *Ecosystems and Human Well-being: Synthesis* (Washington, DC: Island Press) and European Communityies (2008) *The Economics of Ecosystems and Biodiversity*, TEEB: an interim report.

23 Wilkinson C (2008) in *The Economics of Ecosystems and Biodiversity report and Millennium Ecosystem Assessment*, 2005.

24 Malhi Y, Wood D, Baker T, Wright J, Phillips O, Cochrane T, Meir P, Chave J, Almeida S, Arroyo L, Higuchi N, Killeen T, Laurance S, Laurance W, Lewis S, Monteagudo A, Neill D, Núñez Vargas P, Pitman N, Quesada C, Salomão R, Silva J, Lezama A, Terborgh J, Martinez R and Vinceti (2006) 'The regional variation of aboveground live biomass in old-growth Amazonian forests' *Global Change Biology* **12**(7): 1107–1138.

25 Rial J, Pielke R Sr, Beniston M, Claussen M, Canadell J, Cox P, Held H, De Noblet-Ducoudré N, Prinn R, Reynolds J and Salas J (2004) Non linearities, feedbacks and critical thresholds within the earth's climate system *Climatic Change* **65**: 11–38.

26 Lenton T, Held H, Krieglar E, Hall J, Lucht W, Rahmstorf S, Schellnhuber HJ (2008) Tipping elements in the Earth's climate system *Proceedings of the National Academy of Sciences* **105**(6): 1786–1793.

27 Kirchoff S (2008) 'Greenspan takes one on the chin, admits flaws in the system,' available at: http://abcnews.go.com/Business/story?id=6100290&page=1

28 Kasser T(2002) *The high price of materialism* (London: MIT Press).

29 Wilkinson RG and Pickett KE (2009) *The spirit level: why more equal societies almost always do better* (London: Allen Lane).

30 WEO projections are also used by the government's Committee on Climate Change (CCC).

31 Hirsch D, Davis A and Smith N (2009) *A minimum income standard for Britain in 2009* (York: JRF).

32 Wilkinson and Pickett (2009) *op. cit.*

33 350 parts per million is what many scientists, climate experts, and progressive national governments are now saying is the safe upper limit for CO_2 in our atmosphere. A major global campaign, www.350.org, lobbied for this target to be adopted at Copenhagen.

34 Meyer A (2001) *Contraction & convergence: The global solution to climate change*, Schumacher Briefings 5, Green Books on behalf of the Schumacher Society.

35 The term was originally coined by the American economist Neva Goodwin.

36 SROI is an adjusted form of cost-benefit analysis that captures value created for multiple stakeholders and across a triple bottom line. The UK Government recently published a guide to using SROI: Cabinet Office, Office of the Third Sector, (2009) *Social Return on Investment: An introduction.* Available at: http://www.sroi-uk.org/content/view/5/65/

37 Amartya Sen, (1999) *Development as Freedom* (New York: Random House).

38 Policies for distributing income, time and carbon are currently being developed in separate silos. While we focus on income and time in this report, **nef** is also working on incorporating carbon into this redistribution process. The ways in which people use income, time and carbon are often closely related and interdependent. This has implications for social justice as well as for sustainable development. Through research and dialogue with experts in each of these fields, we are seeking to identify opportunities for achieving better outcomes for sustainable social justice through a more integrated approach to the distribution of income, time and carbon.

39 Davis J, Sandstrom S, Shorrocks A and Wolff E (2008) *The World Distribution of Household Wealth, World Institute for Development Economics Research Discussion Paper* No. 2008/03 (Geneva: United Nations University).

40 FAO (1990) *Report on the 1990 World Census of Agriculture, Food and Agriculture Organization of the United Nations.* Available at: http://www.fao.org/economic/ess/world-census-of-agriculture/report-on-the-1990-world-census-of-agriculture/en [22 September 2009].

41 Wilkinson RG and Pickett KE (2007) 'The problems of relative deprivation: why some societies do better than others' *Social Science and Medicine* **65(9):** 1965–1978.

42 Personal wealth in the UK is approximately £3.3 trillion. Given the numbers of deaths and the number of births (based on the latest ONS figures) and assuming that in any year the wealth distribution of the deceased reflects that of society, then an inheritance levied on all estates at 67 per cent would be sufficient to generate between £40,000 and £50,000 to fund the Citizen's Endowment. Clearly, there is no reason to assume that the amounts needed each year to fund the Endowment would tally with revenues from inheritance. However, over the long-term this would be the case, so that inheritance taxes would be paid into a dedicated Endowment Fund which could hold revenues and ensure sufficient remained in any given year to pay the Endowments. As some of these assets could be redistributed communally, the level of the endowment would be less tha£40.000–£50,000. If we were to imagine a 50–50 split between Citizen's Endowments and Community Endowments, this would lead to an annual Citizen's Endowment of around £20,000–£25,000, but also regular Community Endowments at the same per capita level allocated equitably across the country.

43 CLTs have been particularly successful in the United States and in rural areas of the UK and Scotland where land has been acquired by philanthropic bodies or landlords. Letchworth Garden city was developed in 1903 via a co-operative land society mechanism. Scotland has a rural history of absentee landlords and consequent neglect, and has therefore led the way towards new kinds of community asset-holding. The 'community right to buy' was enshrined in the Land Reform (Scotland) Act 2003, and the parallel investment through the Scottish Land Fund was made possible in 2000, with initial capitalisation from the UK National Lottery's New Opportunities Fund. Thanks to those two interventions, a growing number of large estates and islands have been bought by community trusts. By June 2005, 200 communities had been helped and £12 million committed to eligible projects.

44 Named after German economist Rudolf Meidner, the 'plan' was implemented in 1982 by the Social Democratic government. It required 20 per cent of large company profits to be paid into regional mutual funds. Controlled by workers and consumers, these groups gained representation on company boards as their holdings grew. The Meidner Plan was to fizzle out without achieving any lasting impact, not least as the government was lukewarm about it and what enthusiasm there was within the Social Democrats evaporated after the financial crisis of 1992. If pursued steadily, however, there is no reason why such a model could not be successfully implemented.

45 See Burczak TA (2006) *Socialism after Hayek* (Michigan: University of Michigan Press) for an excellent rationale for such a system.

46 See Bacup (2008) for an account of how the ownership transfer process could happen in practice.

47 See Krugman P (2008) *The myth of Asia's miracle*, for an interesting skit on this in the context of the Asian financial crisis of 1997/1998. Available at: http://www.foreignaffairs.com/articles/50550/paul-krugman/the-myth-of-asias-miracle#

48 Burczak, T (2006) *op cit.*

49 The term 'triple bottom line' was coined by John Elkington, founder of *Sustainability* and referred to the need for companies to consider social and environmental 'returns' as well as the commercial variety.

50 In economics, an 'externality' of an economic transaction is an impact on a party that is not directly involved in the transaction. In such a case, prices do not reflect the full costs or benefits to all those affected, and to society as a whole, of the production or consumption of a particular product or service. An advantageous impact is called a 'positive externality', while a detrimental impact is called a 'negative externality'. Producers and consumers in a market may either not bear all of the costs or not reap all of the benefits of the economic activity. For example, manufacturing that causes air pollution imposes costs on the whole society, while fire-proofing a home improves the fire safety of neighbours.

51 Samuelson PA (1954). 'The pure theory of public expenditure' *Review of Economics and Statistics* **36:** 387–389.

52 This kind of coaching approach has been particularly effective in supporting small enterprises in areas experience economic disadvantage in the UK. **nef**'s BizFizz program turns the passion and enthusiasm of individuals into a driving force by mobilising the skills and resources within communities. BizFizz locates a coach in the heart of the community who gets to know that community thoroughly, via long-term relationships with local people, clubs, pubs, societies, schools and faith groups. Unlike conventional business support agencies their role is not to sit on committees, dole out funding, training or provide premises but simply to support entrepreneurs by helping them to follow their passions. What is important in this model is that, rather than being dictated to by a centrally imposed 'expert', the clients make their own choices and then advice is offered according to those choices. The coach never initiates or motivates but simply responds to the clients' requests for support; the advice is based on trust, rapport and the coach's knowledge of the individual and local context. www.bizfizz.org.uk

53 Such an approach was proposed in Thaler R and Sunstain C (2008) *Nudge: improving descisions about health, wealth and happiness* (Yale, Yale Univerisity Press) which drew heavily on the work of Nobel Laureate Daniel Kahneman amongst other authors.

54 Barr MS, Mullainathan S and Shafr E (2009) 'The case for behaviorally informed regulation', in Moss D (ed.) *New Perspectives on Regulation* (Harvard: Harvard Business School) http://www.tobinproject.org/twobooks/pdf/New_Perspectives_Full_Text.pdf

55 Coote A and Franklyn J (2009) *Green well fair: Three economies for social justice* (London: **nef**).

56 Stephens L, Ryan-Collins J and Boyle D (2008) *Co-production: A manifesto for growing the core economy* (London: **nef**).

57 Coote and Franklin (2009) *op. cit.*

58 Bergstroma T and Sören B (1996).'The political economy of subsidized day care' *European Journal of Political Economy* **12:** 443-457.

59 Simms A (2009) *9 meals from anarchy* (London: **nef**). Available at: http://www.neweconomics.org/gen/uploads/55ksx4awtkarwz55oc4tek5526112008122020.pdf

60 Benjamin A (2008) 'Hurtling towards no town Britain: Support local shops or towns will disappear' *The Guardian* Available at: http://www.guardian.co.uk/society/joepublic/2008/sep/03/communities

61 Walker, B., C. S. Holling, S. R. Carpenter, and A. Kinzig. 2004. Resilience, adaptability and transformability in social–ecological systems. *Ecology and Society* 9(2): 5. [online] URL: http://www.ecologyandsociety.org/vol9/iss2/art5/

62 Lietaer B, Ulanowicz R and Goerner S (2008) *White Paper on all the options for managing a systemic bank crisis.* Available at: www.er.ethz.ch/inspire/systemic_bank_crises

63 *Personnel Today,* Redundancy Tracker, http://www.personneltoday.com/articles/2009/04/08/48935/redundancy-tracker-and-uk-job-losses.html; Stuart Rose, Chief Executive of Marks and Spencer, said following the job losses on 7 January 2009: 'We've got 25 underperforming Simply food stores... which are simply too small for us to be able to deal with in a profitable way and we've had to take some harsh action on that today.'

64 Ricardo, D., (1817) *On the Principles of Political Economy and Taxation*, [online] http://www.econlib.org/library/Ricardo/ricP2a.html#Ch.7,%20On%20Foreign%20Trade

65 uktradeinfo.com (2008); see **nef**'s interdependence reports for more information: www.neweconomics.org/interdependence

66 Daly H (2008) 'A steady state economy', submission to the Sustainable Development Commission, UK.

67 Veblen T (1899) *The theory of the leisure class: an economic study of institutions* (Montana, USA: Kessinger Publishing).

68 Adapted from Ziman J (2003) 'Subsidiarity' in Boyle D and Conisbee M (2003), *Return to scale: alternatives to globalisation* London: **nef** and the Institute of Contemporary Arts).

69 Sacks J (2005) *Public spending for public benefit* (London: **nef**).

70 Jacobs J (1984) *Cities and the wealth of nations: Principles of economic life* (New York: Vintage) Chapter 11:156–181 suggests the City (or City-region) is a much more salient economic entity than the state and emerge naturally through import substitution.

71 On mutual credit networks, see Greco T (2009) *The end of money and the future of civilisation* (White River Junction, Vermont: Chelsea Green Publishing) Chapters 12 and 16.

72 According to the International Reciprocal Trade Association (IRTA, the leading trade association for what is now being called the commercial barter industry), in 2007 the industry enabled trades worth more than US$10 billion – a figure growing at an estimated annual rate of 8 per cent. IRTA estimates that 400,000 business firms – mostly SMEs – use the services of commercial 'barter' companies. www.irta.org

73 Goetz SJ and Rupasingha A (2006), 'Wal-mart and social capital' *American Journal of Agricultural Economics* **88:** 5. Quoted in Boyle D and Simms A (2009) *The new economics: A bigger picture* (London: Earthscan).

74 Goetz and Rupasingha, (2006) *op. cit.*

75 Archon F and Olin Wright E (2003) *Deepening democracy: institutional innovations in empowered participatory governance* (London and New York: Verso).

76 Sennett R (2008) *The craftsman* (London: Penguin Books).

77 Fromm E (1997) *To have or to be* (London: Continuum).

78 *Ibid.*

79 HIC (2005) Sustainable urban district Freiburg-Vauban (Freiburg: Habitat International Coalition).

80 http://www.wwcd.org/policy/US/newdeal.html#END

81 Poulter) (2008) 'Sales of bread-makers boom as price of a loaf rises by 20 per cent in a year' *The Daily Mail* 28 May.

82 Spedding A (2009) *New Blood: Attracting the best young people to agriculture* (London: Royal Agricultural Society of England).

83 Perez R, Cuban Agriculture expert.

84 Burton J and Hubacek K (2007) 'Is small beautiful? A multi-criteria assessment of small-scale energy technology applications in local governments' *Energy Policy* **35**: 6402–6412.

85 http://www.energy4all.co.uk/energy_home.asp [7 January 2009].

86 Conaty P *et al* (2009) *Decarbonising Local Economies* (London: **nef).**

87 Land Value Taxation is a method of raising public revenue by means of an annual tax on the rental value of land. Land (unlike goods and services) has no cost of production and the supply of land is inelastic, hence a tax should not have distortive effects on the economy or prices (unlike taxes on labour for example). For a strong argument for its implementation in the UK as a means of reducing the volatility of the housing market and making homes more affordable, see Lloyd T (2009) *Don't bet the house on it* (London: Compass). Available at: http://clients.squareeye.com/uploads/compass/documents/Compass%20Housing%20web.pdf. Llyod estimates the lack of tax on Capital gains created a subsidy worth about £13 billion in 2007 (p.16).

88 Where populations were relatively sparse, there might be a need to augment locally raised funds with national public monies to ensure the ability to provide high quality health and education services was uniformly high in all areas of the country.

89 Feed-in-tariffs are incentives to encourage the adoption of renewable energy technologies, such as solar. A household, or organisation, adopting the technology is paid a guaranteed price for providing surplus energy to the grid. Feed-in-tariff schemes have been adopted my many countries, and vary from context to context.

90 Fractional-reserve banking is the banking practice in which banks keep only a fraction of their deposits in reserve (as cash and other highly liquid assets) and lend out the remainder, while maintaining the simultaneous obligation to redeem all these deposits upon demand. Fractional reserve banking necessarily occurs when banks lend out any fraction of the funds received from deposit accounts. This practice is universal in modern banking. Golin J (2001) *The bank credit analysis kandbook: A guide for analysts, bankers and investors* (Chichester: John Wiley & Sons).

91 Simms A, Johnson V, Smith J and Mitchell S (2009) *The consumption explosion: The 3rd UK interdependence report* (London: **nef)** p. 20.

92 Crooks E and Wallis W (2007) 'Africa aid wiped out by rising cost of oil' *Financial Times* 28 December.

93 Jackson T (2009) *Prosperity without growth* (London: Sustainable Development Commission).

94 Daly H (1996) *Beyond growth: the economics of sustainable development* (Boston: Beacon Press), p. 28.

95 Victor P A (2008) *Managing without Growth: slower by design, not disaster* (Cheltenham: Edward Elgar).

96 Hansen J, Sato M, Kharecha P, Beerling D, Masson-Delmotte V, Pagani M, Raymo M, Royer D and Zachos J (2008) 'Target atmospheric CO_2: where should humanity aim?' *Open Atmospheric Science Journal* **2**: 217–231.

97 *Ibid.*

98 Methane (CH_4), Nitrous Oxide (N_2O), hydroflurocarbons (HCFCs), perflurocarbons (PFCs) and sulphur hexafluoride (SF_6).

99 Global warming potential is defined by the IPCC as the ratio of the time-integrated radiative forcing from the instantaneous release of 1 kg of a trace substance relative to that of 1kg of a reference gas.

100 Anderson and Bows (2008) *op. cit.*

101 Meinshausen M, Meinshausen N, Hare W, Raper S, Frieler K, Knutti R, Frame D and Allen M (2009) 'Greenhouse-gas emissions targets for limiting global warming to 2 °C *Nature* **458**: 1158–1162.

102 Allen M, Frame D, Huntingford C, Jones C, Lowe J, Meinshausen M and Meinshausen N (2009) 'Warming caused by cumulative carbon emissions towards the trillionth tonne' *Nature* **458**: 1163–1166.

103 Meinshausen *et al.* (2009) *op. cit.*

104 Allen *et al.* (2009) *op. cit.*

105 Meinshausen *et al.* (2009) *op. cit.*

106 Allen *et al* (2009) *op. cit.*

107 Johnson and Simms (2008) *100 Months: Technical note op. cit.*

108 Anderson and Bows (2008) *op. cit.*

109 http://www.oecd.org/document/24/0,3343en_2649_34637_2671576_1_1_1_1,00.html#data

110 Godfrey *et al* (2002) Estimating the costs of 'not being in education, training, or employment' at age 16-18. (York: Social Policy Research Unit).

111 The Prince's Trust (2007) The costs of social exclusion: Counting the cost of youth disadvantage in the UK. (London: Prince's Trust).

112 Godfrey *et al* (2002) *op. cit.*

113 Swanton K (2008) Healthy weight, healthy lives: A toolkit for developing local strategies (London: Department of Health) p. 29.

114 Brand S and Price R (2000) Economic and social costs of crime (London: Home Office.)

115 OECD (2009) OECD Fact book 2009: Economic, environmental and social statistics. Available at: http://statlinks.oecdcode.org/302009011P1T135.XLS [7 August 2009].

116 Prince's Trust (2007) *op. cit.*

117 Godfrey *et al* (2002) *op. cit.*

118 *Ibid.*

119 European Monitoring Centre for Drugs and Drug Addiction (2005) http://www.emcdda.europa.eu/stats07/pdutab01a [9 August 2009].

120 Home Office (2000) The economic and social costs of class A drug use in England and Wales (London: Home Office).

121 Drugscope http://www.publications.parliament.uk/pa/cm200102/cmselect/cmhaff/318/31804.htm#note41

122 The European Opinion Research Group (2003) The mental health status of the European population Eurobarometer 58.2 (Brussels: The European Opinion Research Group). Available at: http://ec.europa.eu/health/ph_determinants/life_style/mental_eurobaro.pdf [9 August 2009].

123 This did not include Norway, which was taken from the Statistics Norway for that year. Statistics Norway http://www.ssb.no/helsetilstand_en/ [9 August 2009].

124 The European Opinion Research Group (2003) *op. cit.* p1.

125 McCrone P, Dhanasiri S, Patel A, Knapp M, Lawton-Smith S (2008) *Paying the Price: The cost of mental health care in Britain until 2026* (London: King's Fund).

126 *Ibid.*

127 Family Matters Institute (no date) The cost of family breakdown, a report by Family. Available at: http://www.familymatters.org.uk/published.php [9 August 2009].

128 HM Treasury (2009) Public Expenditure Statistical Analyses (PESA). Based on current expenditure on services by function using economic affairs sub-category: enterprise and economic development.

129 International Energy Agency (2008) *World Energy Outlook 2008 Edition* (Paris: IEA).

130 Anderson K, Bows A and Mander S (2008) 'From long-term targets to cumulative emission pathways: Reframing UK climate policy' *Energy Policy* **36**(10): 3714-3722.

131 Hansen, J et al (2008) 'Target Atmospheric CO2: Where should humanity aim?' *The Open Atmospheric Science Journa*l **2**: 217-231.

132 Population Division of the Department of Economic and Social Affairs of the United Nations Secretariat, World Population Prospects: The 2008 Revision. Available at http://esa.un.org/unpp [Last accessed September 6, 2009].

133 Anderson *et al.* (2006) Decarbonising Modern Societies: integrated scenarios process and workshops, Tyndall Centre for Climate Change Research Technical Report 48, available at: http://www.tyndall.ac.uk/research/theme2/final_reports/t3_24.pdf

134 Ackerman F, Stanton E, Hope C, Alberth S (2009) 'Did the Stern Review underestimate US and global climate damages?' *Energy Policy* **37**:2717–2721.

135 http://www.hm-treasury.gov.uk/d/climatechnage_judge.pdf

136 http://climatechange.cbi.org.uk/uploaded/climatereport2007mckinseyapp.pdf

137 van Vurren D and Riahi K (2008) Do recent emission trends imply higher emissions forever? *Climatic Change* **91**: 237-248.

138 Department of Trade and Industry (2007) *Meeting the Energy Challenge. A White Paper on Energy* (Based on calculations on page 329 and costs as share of GDP from their model).

139 http://www.ukerc.ac.uk/Downloads/PDF/07/0706ESMMARKALpresNS.pdf

In choosing the title *The Great Transition* we took and adapted the title of Karl Polanyi's work, *The Great Transformation*. But we'd also like to acknowledge Kenneth Boulding, economist (1910–1993). Our research suggests he first used the term in a speech, *The Horizons of Economic Knowledge and the Great Transition*, presented at Carroll College, Waukesha, Wisconsin, on January 15, 1963. If we looked hard we would probably find an even earlier reference. The Global Scenario Group (GSG), an international body of scientists convened in 1995 by the Tellus Institute and Stockholm Environment Institute to examine the requirements for a transition to a sustainable global society also use the phrase in their essay *Great Transition: the Promise and Lure of the Times Ahead* and subsequent papers on the theme. Then, there is also the international Transition Town movement which is seeking to apply many of these ideas in a practical way. It is thought to be the fastest growing social movement in the UK. This all goes to show that there's rarely such a thing as a truly new idea, and great minds often think alike...